Collins

ROYAL
OBSERVATORY
GREENWICH

Night Sky
ALMANAC

A STARGAZER'S GUIDE TO

2024

Storm Dunlop & Wil Tirion

Published by Collins
An imprint of HarperCollins Publishers
Westerhill Road, Bishopbriggs, Glasgow, G64 2QT
www.harpercollins.co.uk

HarperCollins Publishers
Macken House, 39/40 Mayor Street Upper, Dublin 1, D01 C9W8, Ireland

In association with
Royal Museums Greenwich, the group name for the National Maritime Museum,
Royal Observatory Greenwich, the Queen's House and *Cutty Sark*
www.rmg.co.uk

A catalogue record for this book is available from the British Library

ISBN 978-0-00-860429-5

10 9 8 7 6 5 4 3 2 1

Printed in the UK using 100% Renewable Electricity at CPI Group (UK) Ltd

If you would like to comment on any aspect of this book, please contact us at the above
address or online.
e-mail: collins.reference@harpercollins.co.uk

facebook.com/CollinsAstronomy
@CollinsAstro

Contents

Data used in the Night Sky Almanac

The data given in this almanac, such as timings and distances between objects, have been computed with a program developed by the US Naval Observatory in Washington DC (MICA, the Multiyear Interactive Computer Almanac), widely regarded as the most accurate computation. As such, the data may differ slightly from information given elsewhere.

Introduction

The aim of this book is to help people to find their way around
the night sky and to understand what is visible every month,
from anywhere in the world. The stars that may be seen depend
on where you are on Earth, but even if you travel widely,
this book will show you what you can see. The night sky also
changes from month to month and these changes, together
with some of the significant events that occur during the year
are described and illustrated.

The charts that are used differ considerably from those
found in most astronomy books, and have been specifically
designed for use anywhere in the world. A full description of
how to use and understand the monthly charts is given on
pages 40–43.

Sunrise, sunset and twilight

The conditions for observing naturally vary over the course of
the year and one's location on Earth. Sunrise and sunset vary
considerably, depending in particular on one's latitude. Sunrise
and sunset times are given each month for nine different
locations around the world. These places are shown in a **bold**
typeface on the world map on pages 44–45. Sunrise and sunset
times are given for the first and last days in every month, for
these specific locations. Another factor that influences what
may be seen is twilight at dusk and dawn. Again, this varies
considerably with one's latitude on Earth. The diagrams on
pages 251–253 show how this varies for the nine locations, which
have been chosen to show the range of variation, rather than
just for the importance of the places that have been included.
The different stages of twilight and how they affect observing
are also explained there.

Moonlight

Yet another factor that affects the visibility of objects is the
amount of moonlight in the sky. At Full Moon, it may be very
difficult to see some of the fainter stars and objects, and even
when the Moon is at a smaller phase it seriously interferes
with visibility if it is near the stars or planets in which you are
interested. A full lunar calendar is given for each month and

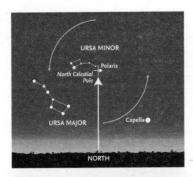

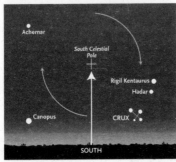

The altitude of the Celestial Pole equals the observer's latitude.

may be used to see when nights are likely to be darkest and best for observing.

The celestial sphere

All the objects in the sky (including the Sun, Moon, and stars) appear to lie at some indeterminate distance on a large sphere, centred on the Earth. This celestial sphere has various reference points and features that are related to those of the Earth. If the Earth's rotational axis is extended, for example, it points to the North and South Celestial Poles, which are thus in line with the North and South Poles on Earth. As shown in the diagrams, the altitude of the celestial pole is equal to the observer's latitude, whether in the north or south. Similarly, the celestial equator lies in the same plane as the Earth's equator, and divides the sky into northern and southern hemispheres.

It is useful to know some of the special terms for various parts of the sky. As seen by an observer, half of the celestial sphere is invisible, below the horizon. The point directly overhead is known as the **zenith**, and this point is shown on the monthly charts for several different latitudes, where it is an important reference point. The (invisible) point below one's feet is the **nadir**. The line running from the north point on the horizon, up through the zenith and then down to the south point is the **meridian**. This is an important invisible line in the sky, because objects are highest in the sky, and thus easiest to see, when they cross the meridian in the south. Objects are said to transit, when they cross this line in the sky.

In this book, reference is sometimes made in the text and in the diagrams to the standard compass points around the horizon. The position of any object in the sky may be described by its **altitude** (measured in degrees above the horizon) and its **azimuth** (measured in degrees from north, 0°, through east, 90°, south, 180°, and west, 270°). Experienced amateurs and professional astronomers also use another system of specifying locations on the celestial sphere, but that need not concern us here, where the simpler method will suffice.

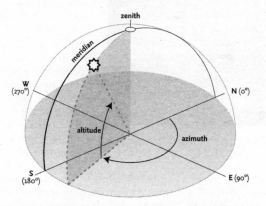

Measuring altitude and azimuth on the celestial sphere.

The celestial sphere appears to rotate about an invisible axis, running between the north and south celestial poles. The location (i.e., the altitude) of the celestial poles depends entirely on the observer's position on Earth or, more specifically, their latitude.

Right ascension and declination

On the previous pages we have mentioned how a simple method, involving altitude and azimuth, measured relative to the observer's horizon, may be used to specify the position of an object in the sky. Astronomers, however, use another, more precise method, which does not depend on the observer's position on Earth (and thus on their local horizon). This involves the two co-ordinates, *right ascension* (RA) and *declination* (dec). Right ascension is measured eastwards (to the left) from the *First Point of Aries* (page 82) in hours and minutes of time (and very occasionally, in seconds) or else (less frequently) in degrees. Because the Earth rotates once in 24 hours, one hour of right ascension equals 15 angular degrees. The sky appears to rotate by this amount in one hour.

All objects in the sky appear to be located on an imaginary sphere: the celestial sphere. There are, however, certain fixed points on the celestial sphere, related to points on the Earth. The North Celestial Pole (NCP) and the South Celestial Pole (SCP) are located in line with the projection of the Earth's rotational axis onto that sphere. In the north, the NCP is very close to Polaris, which has been known as the North Star since antiquity. In a similar way, the celestial equator is the projection onto the sphere of the Earth's equator. The second co-ordinate, declination, is simply the angular distance, in degrees, north or south of the celestial equator. The Sun has a declination of zero when it appears to cross the celestial equator at the equinoxes.

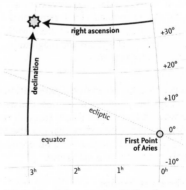

Measuring right ascension (RA) and declination (dec) on the celestial sphere.

The ecliptic and the zodiac

Another important line on the celestial sphere is the Sun's apparent path against the background stars – in reality the result of the Earth's orbit around the Sun. This is known as the *ecliptic*. The point where the Sun, apparently moving along the ecliptic, crosses the celestial equator from south to north is known as the vernal (or northern spring) equinox, which occurs around March 21. At this time (and at the northern autumnal equinox, on September 22 or 23, when the Sun crosses the celestial equator from north to south) day and night are almost exactly equal in length. (There is a slight difference, but that need not concern us here.) The vernal equinox is currently located in the constellation of Pisces, and is important in astronomy because it defines the zero point for a system of celestial coordinates, which is, however, not used in this book.

The Moon and planets are to be found in a band of sky that extends 8° on either side of the ecliptic. This is because the orbits of the Moon and planets are inclined at various angles to the ecliptic (i.e., to the plane of the Earth's orbit). This band of sky is known as the zodiac, and when originally devised, consisted of twelve *constellations*, all of which were considered to be exactly 30° wide. When the constellation boundaries were

Most of the brightest stars have names officially recognized by the International Astronomical Union. A list of these, with their Bayer designations and magnitudes, is given on pages 260–261.

formally established by the International Astronomical Union in 1930, the exact extent of most constellations was altered, and nowadays the ecliptic passes through thirteen constellations. Because of the boundary changes, the Moon and planets may actually pass through several other constellations that are adjacent to the original twelve.

The constellations

In the western astronomical tradition, the celestial sphere has always been divided into various constellations, most dating back to antiquity and usually associated with certain myths or legendary people and animals. Nowadays, 88 constellations cover the whole sky, and their boundaries have been fixed by international agreement. Their names (in Latin) are largely derived from Greek or Roman originals. (A full list of the constellations is given on pages 256–258, with their names in English, their abbreviations, and genitive forms.) Some of the names of the most prominent stars are of Greek or Roman origin, but many are derived from Arabic names. Some bright stars have no individual names, and for many years, they were identified by terms such as 'the star in Hercules' right foot'. A more sensible scheme was introduced by the German astronomer Johannes Bayer in the early seventeenth century. Following his scheme – which is still used today – most of the brightest stars are identified by a Greek letter followed by the genitive form of the constellation's Latin name. An example is the Pole Star, also known as Polaris and α Ursae Minoris. The Greek alphabet is shown on page 258.

Asterisms
Apart from the constellations, certain groups of stars, which may form a small part of a larger constellation, are readily recognizable and have been given individual names. These groups are known as asterisms, and the most famous (and well-known) is the 'Plough' or 'Big Dipper', the common name for the seven brightest stars in the constellation of Ursa Major, the Great Bear. The names and identifications of some popular asterisms are given in the list on page 259.

The Moon
As it passes across the sky from west to east in its orbit around the Earth, the Moon moves by approximately its diameter (about half a degree) in an hour. Normally, in its orbit, the Moon passes above or below the direct line between Earth and Sun (at New Moon) or outside the area obscured by the Earth's shadow (at Full Moon). Occasionally, however, the three bodies are more-or-less perfectly aligned to give an eclipse: a solar eclipse at New Moon, or a lunar eclipse at Full Moon.

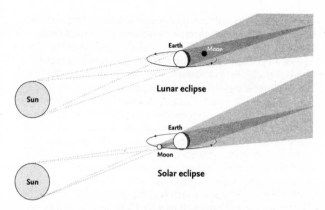

When the Moon passes through the Earth's shadow (top), a lunar eclipse occurs. When it passes in front of the Sun (below) a solar eclipse occurs.

Depending on the exact circumstances, a solar eclipse may be merely partial (when the Moon does not cover the whole of the Sun's disc); annular (when the Moon is too far from Earth in its orbit to appear large enough to hide the whole of the Sun); or total. Total and annular eclipses are visible from very restricted areas of the Earth, but partial eclipses are normally visible over a wider area. Two forms of solar eclipse occur this year, and are described in detail in the appropriate month.

Precautions must always be taken when viewing even partial phases of a solar eclipse to avoid damage to your eyes. Only ever use proper eclipse glasses, or a proper solar filter over the full objective of a telescope. The glass 'solar filters' sometimes provided with cheap telescopes should never be used. They are unsafe.

Somewhat similarly, at a lunar eclipse, the Moon may pass through the outer zone of the Earth's shadow, the penumbra (in a penumbral eclipse, which is not generally perceptible to the naked eye); pass so that just part of the Moon is within the darkest part of the Earth's shadow, the umbra (in a partial eclipse); or completely within the umbra (in a total eclipse).

Unlike solar eclipses, lunar eclipses are visible from large areas of the Earth. Again, these are described in detail in the relevant month.

Occasionally, as it moves across the sky, the Moon passes between the Earth and individual planets or distant stars, giving rise to an **occultation**. As with solar eclipses, such occultations are visible from restricted areas of the world, but certain significant occultations are described in detail.

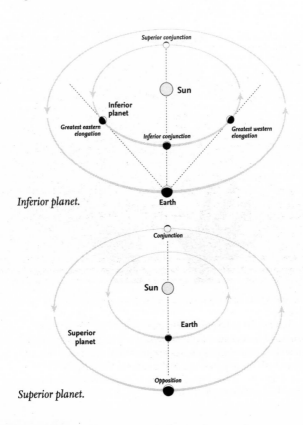

Inferior planet.

Superior planet.

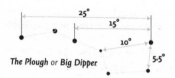

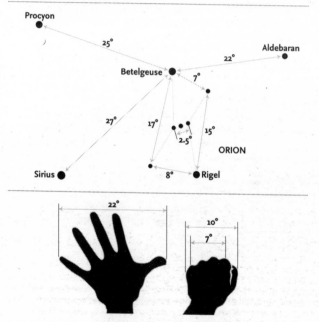

Measuring angles in the sky.

It is often useful to be able to estimate angles on the sky, and approximate values may be obtained by holding one hand at arm's length. The various angles are shown in the diagram, together with the separations of the various stars in the asterism, known as the Plough or Big Dipper, and also for stars around the constellation of Orion.

13

The planets

Because the planets are always moving against the background stars, they are treated in some detail in the monthly pages and information is given when they are close to other planets, the Moon, or any of five bright stars that lie near the ecliptic. Such events are known as ***appulses*** or, more frequently, as ***conjunctions.*** (There are technical differences in the way these terms are defined – and should be used – in astronomy, but these need not concern us here.)

The term conjunction is also used when a planet is either directly behind or in front of the Sun, as seen from Earth. (Under normal circumstances it will then be invisible.) The conditions of most favourable visibility depend on whether the planet is one of the two known as ***inferior planets*** (Mercury and Venus) or one of the three ***superior planets*** (Mars, Jupiter and Saturn) that are covered in detail. Brief details of the fainter superior planets, Uranus and Neptune, are given, especially when they come to opposition.

The inferior planets are most readily seen at eastern or western ***elongation***, when their angular distance from the Sun is greatest. For superior planets and minor planets, they are best seen at ***opposition***, when they are directly opposite the Sun in the sky, and cross the meridian at local midnight.

Events

A number of interesting events are shown in diagrams for each month. They involve the planets and the Moon, sometimes showing them in relation to specific stars. Events have been chosen as they will appear from one of three different locations: from London; from the central region of the USA; or from Sydney in Australia. Naturally, these events are visible from other locations, but the appearance of the objects on the sky will differ slightly from the diagrams. A list of major astronomical events in 2024 is given on pages 24–25.

Meteors

At some time or other, nearly everyone has seen a ***meteor*** – a 'shooting star' – as it flashed across the sky. The particles that

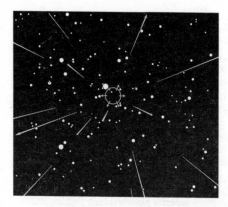

*Meteor shower
(showing the April
Lyrid radiant).*

cause meteors – known technically as 'meteoroids' – range in
size from that of a grain of sand (or even smaller) to the size of
a pea. On any night of the year there are occasional meteors,
known as **sporadics**, that may travel in any direction. These
occur at a rate that is normally between 3 and 8 in an hour. Far
more important, however, are **meteor showers**, which occur at
fixed periods of the year, when the Earth encounters a trail of
particles left behind by a comet or, very occasionally, by a minor
planet (asteroid). Meteors always appear to diverge from a single
point on the sky, known as the **radiant**, and the radiants of
major showers are shown on the charts.

Meteors that come from a circular area, 8° in diameter,
around the radiant are classed as belonging to the particular
shower. All others that do not come from that area are sporadics
(or, occasionally, from another shower that is active at the same
time). A list of the major meteor showers is given on the next
page.

Although the positions of the various shower radiants are
shown on the charts, looking directly at the radiant is not the
most effective way of seeing meteors. They are most likely to
be noticed if one is looking about 40–45° away from the radiant
position. (This is approximately two hand-spans as shown in
the diagram for measuring angles on page 13.)

Motion of the planets

All the outer planets from Mars to Neptune exhibit periods of retrograde motion, when, instead of their normal, orderly progress across the sky from west to east (known as direct motion), they reverse direction and travel from east to west. This retrograde movement continues for a while and then they reverse direction and resume direct motion.

When only the outer planets from Mars to Saturn were known, retrograde motion was one of the main problems that faced astronomers and astrologers when it was believed that the whole universe was centred on the Earth (in a geocentric universe).

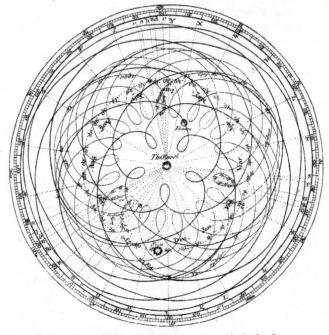

The highly complex pattern of planetary motions that had to be explained on the geocentric model.

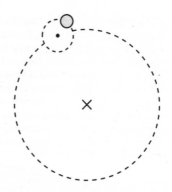

The planet was carried around a small circle, itself carried around the Earth on a larger circle.

The result of this behaviour was a highly complex pattern, which proved difficult to explain on a geocentric model.

Because of the notion that movements in the planetary realm could only occur in 'perfect' circles, the idea was introduced that the planets moved on small circles (epicycles) that were themselves carried around the Earth on circular orbits. This concept was first introduced by Apollonius of Perga, whose dates are unknown, but who lived around 240 to 190 BCE. He studied geometry and astronomy, but most of his writings are lost. The crater Apollonius on the Moon carries his name (see page 19). The various concepts were developed by the Greek astronomer Hipparchus (about 190 to 120 BCE), who also has a lunar crater named after him (see below).

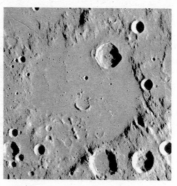

The lunar crater Hipparchus is located near the centre of the Moon.

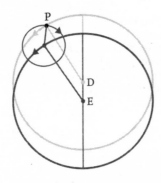

The planet (P) orbited on a small circle, carried round a larger circle, centred at the deferent (D), offset from the Earth (E).

Apart from 'explaining' the retrograde motion of the planets, the epicyclic theory also provided a solution to the apparent changes in the distances of the planets from the Earth.

This way of explaining the motion of the planets, where the circular epicycle was carried around the Earth in a larger circlar orbit prevailed for some years.

But even this idea proved inadequate to describe the motion of the planets, so the concept of the deferent was added. In this the circular motion of the epicycle was carried around the Earth on a circle that was itself not centred on the Earth, but offset from its centre.

The use of the epicycle and deferent was developed and propagated by the great astronomer Ptolemy (approximate dates 100 to 170 CE), who found that he had to introduce further terms, which he denoted

The astronomer Ptolemy, assisted by Urania, the Muse of Astronomy. Ptolemy is shown with a crown because at the time of this image (1508) he was confused with the rulers of Egypt.

the 'eccentric' and the 'equant'. This further complicated the situation, and these terms are not explained here.

All this complexity became redundant, of course, as soon as the view of Copernicus prevailed in which the Earth itself orbited the Sun. It became apparent that the planets displayed retrograde motion when the Earth 'caught up' and 'passed' the planets in their orbits.

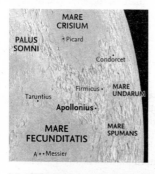

The somewhat indistinct crater Apollonius lies northeast of the Mare Fecunditatis. The western rim is overlain by two smaller craters. There is a flat central floor, flooded by lava.

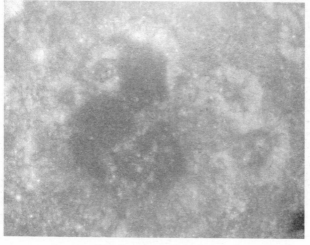

Meteor Showers

Shower	Dates of activity 2024	Date of maximum 2024	Possible hourly rate
Quadrantids	Dec. 28 to Jan. 12	Jan. 3–4	110
α-Centaurids	Jan. 31 to Feb. 20	Feb. 8	6
γ-Normids	Feb. 25 to Mar. 28	Mar. 14–15	6
April Lyrids	Apr. 14–30	Apr. 22–23	18
π-Puppids	Apr. 15–28	Apr. 23–24	var.
η-Aquariids	Apr. 19 to May 28	May 6	50
α-Capricornids	Jul. 3 to Aug. 15	Jul. 30	5
Southern δ-Aquariids	Jul. 12 to Aug. 23	Jul. 30	25
Piscis Austrinids	Jul. 15 to Aug. 10	Jul. 28	5
Perseids	Jul. 17 to Aug. 24	Aug. 12–13	100
α-Aurigids	Aug. 28 to Sep. 5	Sep. 1	6
Southern Taurids	Sep. 10 to Nov. 20	Oct. 10–11	5
Orionids	Oct. 2 to Nov. 7	Oct. 21–22	25
Draconids	Oct. 6–10	Oct. 8–9	10
Northern Taurids	Oct. 20 to Dec. 10	Nov. 12–13	5
Leonids	Nov. 6–30	Nov. 17–18	10
Phoenicids	Nov. 28 to Dec. 9	Dec. 2	var.
Puppid Velids	Dec. 1–15	Dec. 7	10
Geminids	Dec. 4–20	Dec. 14–15	150
Ursids	Dec. 17–26	Dec. 22–23	10

Some Interesting Objects

Messier IC / NGC	Name	Type	Constellation
—	47 Tucanae	globular cluster	Tucana
—	Hyades	open cluster	Taurus
—	Double Cluster	open cluster	Perseus
—	Melotte 11	open cluster	Coma Berenices
M3	—	globular cluster	Canes Venatici
M4	—	globular cluster	Scorpius
M8	Lagoon Nebula	gaseous nebula	Sagittarius
M11	Wild Duck Cluster	open cluster	Scutum
M13	Hercules Cluster	globular cluster	Hercules
M15	—	globular cluster	Pegasus
M22	—	globular cluster	Sagittarius
M27	Dumbbell Nebula	planetary nebula	Vulpecula
M31	Andromeda Galaxy	galaxy	Andromeda
M35	—	open cluster	Gemini
M42	Orion Nebula	gaseous nebula	Orion
M44	Praesepe	open cluster	Cancer
M45	Pleiades	open cluster	Taurus
M57	Ring Nebula	planetary nebula	Lyra
M67	King Cobra Cluster	open cluster	Cancer
IC 2602	Southern Pleiades	open cluster	Carina
NGC 752	—	open cluster	Andromeda
NGC 3242	Ghost of Jupiter	planetary nebula	Hydra
NGC 3372	Eta Carinae Nebula	gaseous nebula	Carina
NGC 4755	Jewel Box	open cluster	Crux
NGC 5139	Omega Centauri	globular cluster	Centaurus

Other objects

Certain other objects may be seen with the naked eye under good conditions. Some were given names in antiquity – Praesepe is one example – but many are known by what are called 'Messier numbers', the numbers in a catalogue of nebulous objects compiled by Charles Messier in the late-eighteenth century. Some, such as the Andromeda Galaxy, M31, and the Orion Nebula, M42, may be seen faintly by the naked eye, but all those given in the list here will benefit from the use of binoculars.

Apart from galaxies, such as M31, which contain thousands of millions of stars, there are also two types of cluster: open clusters, such as M45, the Pleiades, which may consist of a few dozen to some hundreds of stars; and globular clusters, such as M13 in Hercules, which are spherical concentrations of many thousands of stars. One or two gaseous nebulae, consisting of gas illuminated by stars within them are also visible. The Orion Nebula, M42, is one, and is illuminated by

In 1781, **Charles Messier** (26 June 1730 – 12 April 1817) published the final version of his catalogue of 110 nebulous objects and faint star clusters that might be confused with comets. The objects in this catalogue are still know as the Messier objects and are always quoted as 'M' numbers. Some of the most famous are M1, the Crab Nebula; M31, the Andromeda galaxy; M42, the Orion Nebula; and M45, the Pleiades.

the group of four stars, known as the Trapezium, which may be seen within it by using a good pair of binoculars. A list of interesting objects is given on the previous page.

Dates and time

Astronomers, worldwide, use a standardized method of expressing the date and time. This prevents confusion in comparing observations made by different observers. The various elements are given in descending order: year, month (three-letter abbreviation to prevent confusion over using numbers), day, hour, minutes, seconds. (In extreme cases, fractions of minutes or seconds may be used.) The date and time are that on the Greenwich meridian (GMT), and ignore any changes for Summer Time / Daylight Saving Time (DST), and any adjustments for local time at the observer's location. This standard is known as Coordinated Universal Time (UTC). In this book and in many others this is generally given as Universal Time (UT). All times given in this book are in UT.

To avoid problems over the changes involved in moving to and from Summer Time / Daylight Saving Time (and the complications over the beginning and end dates) and also the adjustments for local time, experienced astronomers set a (cheap) watch or clock to Universal Time and keep it that way. Smartphone users may use a simple world clock app, provided they lock it to the time (GMT) on the Greenwich meridian.

Similarly, the date given for an event is the date as it applies at the Greenwich meridian, i.e., in UT. Occasionally, this may differ from the date as given by your local time. An event that occurs (say) late in the night in Europe may seem to occur on the previous day to an observer to the west (such as in the USA), when local time is taken into account. This is another complication that is avoided by using the Universal Time standard.

By 1845, five asteroids had been discovered; by 1868, 100 were known. Numbers increased dramatically following the introduction of photographic methods and dedicated searches and many thousands are now known. (1000 by 1920; 100,000 by 2005 and over one million to date.) The majority of these objects orbit in the inner Solar System, in the Main Belt between Mars and Jupiter, with the Trojan asteroids locked into Jupiter's orbit. There are additional 'families' of asteroids whose orbits take them past the inner planets, Earth, Venus and Mercury. There is also a large population in the outer Solar System, beyond the orbit of Neptune, these objects being known as Trans-Neptunian Objects (TNOs). There are various names for this outer belt of objects, part being known as the Kuiper Belt.

Major Events in 2024

Jan. 03	Earth at perihelion (closest to the Sun)
Jan. 03–04	Quadrantid meteor shower maximum
Jan. 12	Mercury at greatest elongation west
Feb. 08	α-Centaurid meteor shower maximum
Mar. 03	Minor planet (3) Juno at opposition
Mar. 10	Daylight Saving Time (DST) begins in US
Mar. 10	Moon at perigee = 356,896 km (closest of year)
Mar. 14–15	γ-Normid meteor shower maximum
Mar. 24	Mercury at greatest elongation east
Mar. 25	Penumbral lunar eclipse
Mar. 31	Summer Time begins (BST in UK)
Apr. 07	Daylight Saving Time ends (New Zealand & parts of Australia)
Apr. 08	Total solar eclipse (Mexico/USA)
Apr. 22–23	April Lyrid meteor shower maximum
May 06	η Aquariid meteor shower maximum
May 09	Mercury at greatest elongation west
May 19	Minor planet (2) Pallas at opposition
Jul. 06	Dwarf planet (1) Ceres at opposition
Jul. 22	Mercury at greatest elongation east
Jul. 28	Piscis Austrinid meteor shower maximum
Jul. 30	α-Capricornid meteor shower maximum
Jul. 30	Southern δ-Aquariid meteor shower maximum
Aug. 06	Minor planet (7) Iris at opposition
Aug. 12–13	Perseid meteor shower maximum
Sep. 01	α-Aurigid meteor shower maximum

Sep. 05	Mercury at greatest elongation west
Sep. 08	Saturn at opposition
Sep. 21	Neptune at opposition
Sep. 29	Daylight Saving Time begins (New Zealand)
Oct. 02	Annular solar eclipse (Pacific)
Oct. 02	Moon at apogee = 406,516 km (most distant of year)
Oct. 06	Daylight Saving Time begins (parts of Australia)
Oct. 08–09	Draconid meteor shower maximum
Oct. 10–11	Southern Taurid meteor shower maximum
Oct. 21–22	Orionid meteor shower maximum
Oct. 27	Daylight Saving Time ends (UK reverts to GMT)
Nov. 03	Daylight Saving Time ends in US
Nov. 12–13	Northern Taurid meteor shower maximum
Nov. 16	Mercury at greatest elongation east
Nov. 17–18	Leonid meteor shower maximum
Nov. 17	Uranus at opposition
Dec. 02	Phoenicid meteor shower maximum
Dec. 07	Puppid Velid meteor shower maximum
Dec. 07	Jupiter at opposition
Dec. 14–15	Geminid meteor shower maximum
Dec. 14	Minor planet (15) Eunomia at opposition
Dec. 23	Ursid meteor shower maximum
Dec. 25	Mercury at greatest elongation west

The Moon

The Moon at First Quarter.

The Moon

The monthly pages include diagrams showing the *phase* of the Moon (see page 32) for every day of the month, and also indicate the day in the *lunation* (or *age* of the Moon), which begins at New Moon. The diagrams showing the Moon's phase are repeated for southern-hemisphere observers who will see the Moon, south up. Although the main features of the surface – the light highlands and the dark maria (seas) – may be seen with the naked eye, far more features may be detected with the use of binoculars or any telescope. The many craters are best seen when they are close to the *terminator* (the boundary between the illuminated and the non-illuminated areas of the surface), when the Sun rises or sets over any particular region of the Moon and the crater walls or central peaks cast strong shadows. Most features become difficult to see at Full Moon, although this is the best time to see the bright ray systems surrounding certain craters. Accompanying the Moon map on the following pages is a list of prominent features, including the days in the lunation when features are normally close to the terminator and thus easiest to see. A few bright features such as Linné and Proclus, visible when well illuminated, are also listed. One feature, Rupes Recta (the Straight Wall) is readily visible only when it casts a shadow with light from the east, appearing as a light line when illuminated from the opposite direction.

The dates of visibility vary slightly through the effects of *libration*. Because the Moon's orbit is inclined to the Earth's equator and also because it moves in an ellipse, the Moon appears to rock slightly from side to side (and nod up and down). Features near the *limb* (the edge of the Moon) may vary considerably in their location and visibility. (This is easily noticeable with Mare Crisium and the craters Tycho and Plato.) Another effect is that at crescent phases before and after New Moon, the normally non-illuminated portion of the Moon receives a certain amount of light, reflected from the Earth. This *Earthshine* may enable certain bright features (such as Aristarchus, Kepler and Copernicus) to be detected even though they are not illuminated by sunlight.

Moon features

The numbers below indicate the age of the Moon when features are usually best visible.

Abulfeda	6:20	Gassendi	11:25	Philolaus	9:23
Agrippa	7:21	Geminus	3:17	Piccolomini	5:19
Albategnius	7:21	Goclenius	4:18	Pitatus	8:22
Aliacensis	7:21	Grimaldi	13–14:27–28	Pitiscus	5:19
Alphonsus	8:22	Gutenberg	5:19	Plato	8:22
Anaxagoras	9:23	Hercules	5:19	Plinius	6:20
Anaximenes	11:25	Herodotus	11:25	Posidonius	5:19
Archimedes	8:22	Hipparchus	7:21	Proclus	14:18
Aristarchus	11:25	Hommel	5:19	Ptolemaeus	8:22
Aristillus	7:21	Humboldt	3:15	Purbach	8:22
Aristoteles	6:20	Janssen	4:18	Pythagoras	12:26
Arzachel	8:22	Julius Caesar	6:20	Rabbi Levi	6:20
Atlas	4:18	Kepler	10:24	Reinhold	9:23
Autolycus	7:21	Landsberg	10:24	Rima Ariadaeus	6:20
Barrow	7:21	Langrenus	3:17	Rupes Recta	8
Billy	12:26	Letronne	11:25	Saussure	8:22
Birt	8:22	Linné	6	Scheiner	10:24
Blancanus	9:23	Longomontanus	9:23	Schickard	12:26
Bullialdus	9:23	Macrobius	4:18	Sinus Iridum	10:24
Bürg	5:19	Mädler	5:19	Snellius	3:17
Campanus	10:24	Maginus	8:22	Stöfler	7:21
Cassini	7:21	Manilius	7:21	Taruntius	4:18
Catharina	6:20	Mare Crisium	2–3:16–17	Thebit	8:22
Clavius	9:23	Maurolycus	6:20	Theophilus	5:19
Cleomedes	3:17	Mercator	10:24	Timocharis	8:22
Copernicus	9:23	Metius	4:18	Triesnecker	6–7:21
Cyrillus	6:20	Meton	6:20	Tycho	8:22
Delambre	6:20	Mons Pico	8:22	Vallis Alpes	7:21
Deslandres	8:22	Mons Piton	8:22	Vallis Schröteri	11:25
Endymion	3:17	Mons Rümker	12:26	Vlacq	5:19
Eratosthenes	8:22	Montes Alpes	6–8:21	Walther	7:21
Eudoxus	6:20	Montes Apenninus	8	Wargentin	12:27
Fra Mauro	9:23	Orontius	8:22	Werner	7:21
Fracastorius	5:19	Pallas	8:22	Wilhelm	9:23
Franklin	4:18	Petavius	3:17	Zagut	6:20

Map of the Moon

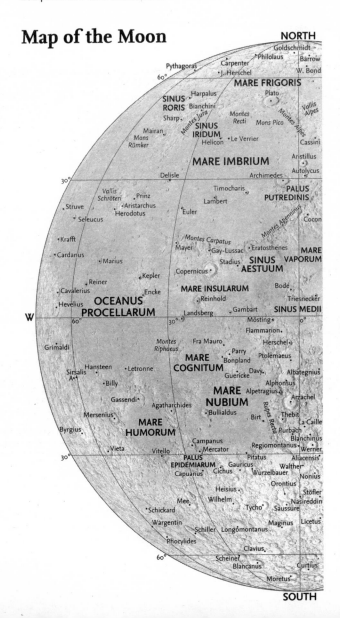

NORTH

Goldschmidt
Philolaus
Carpenter
Barrow
Pythagoras
W. Bond
60°
J. Herschel

MARE FRIGORIS

Harpalus
Plato

SINUS
RORIS
Bianchini
Vallis
Alpes

Sharp
Montes Jura
Montes Recti
Mons Pico
Montes Alpes

Mairan
SINUS
IRIDUM
Helicon
Le Verrier
Cassini

Mons
Rümker

MARE IMBRIUM
Aristillus
Autolycus

Delisle
Archimedes

30°
Timocharis
PALUS
PUTREDINIS

Vallis
Schröteri
Prinz
Lambert

Struve
Aristarchus
Herodotus
Euler
Cocon

Seleucus
Montes Apenninus

Krafft
Montes Carpatus
Eratosthenes
MARE
VAPORUM

Cardanus
Mayer
Gay-Lussac

Marius
Stadius
SINUS
AESTUUM

Reiner
Kepler
Copernicus
Bode

Cavalerius
Encke
MARE INSULARUM
Triesnecker

Hevelius
Reinhold
SINUS MEDII

OCEANUS
PROCELLARUM
Landsberg
Gambart

W
60°
30°
Mösting
0°

Flammarion

Grimaldi
Montes
Riphaeus
Fra Mauro
Herschel

Sirsalis
Hansteen
Letronne
MARE
COGNITUM
Parry
Bonpland
Ptolemaeus

A
Billy
Guericke
Davy
Albategnius

Gassendi
Agatharchides
MARE
NUBIUM
Alphonsus

Mersenius
Reinhold
Alpetragius
Arzachel

Byrgius
Bullialdus
Birt
Thebit
La Caille

Vieta
MARE
HUMORUM
Campanus
Rupes Recta
Purbach

Vitello
Mercator
Regiomontanus
Blanchinus
Werner

30°
PALUS
EPIDEMIARUM
Pitatus
Aliacensis

Capuanus
Gauricus
Cichus
Wurzelbauer
Nonius

Heisius
Orontius
Stöfler

Mee
Wilhelm
Nasireddin

Schickard
Tycho
Saussure
Licetus

Wargentin
Maginus

Schiller
Longomontanus

Phocylides

Clavius
60°
Scheiner
Curtius

Blancanus

Moretus

SOUTH

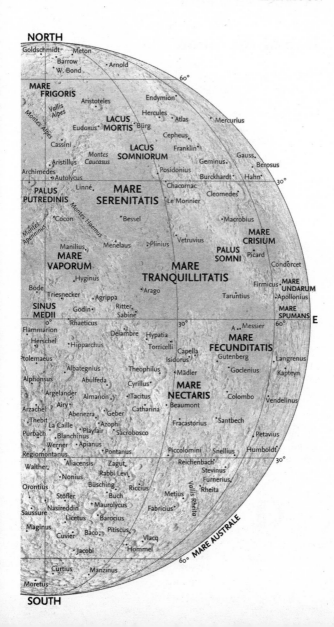

NORTH

Goldschmidt • Meton
Barrow • Arnold
W. Bond
60°
MARE
FRIGORIS
Endymion
Vallis Aristoteles Hercules
Alpes Eudoxus Burg Atlas Mercurius
LACUS Cepheus
MORTIS
Cassini Franklin
Montes LACUS Geminus Gauss
Aristillus Caucasus SOMNIORUM Berosus
Archimedes Posidonius Burckhardt Hahn 30°
• Autolycus
PALUS Linné Chacornac
PUTREDINIS MARE Cleomedes
SERENITATIS Le Monnier
Cocon Montes Haemus Bessel Macrobius
Montes MARE
Apenninus Manilius Menelaus Plinius Vetruvius CRISIUM
MARE PALUS Picard
VAPORUM SOMNI
Hyginus MARE Condorcet
Bode TRANQUILLITATIS Firmicus MARE
Triesnecker Arago UNDARUM
SINUS Agrippa Taruntius Apollonius
MEDII Godin Ritter MARE
0° Sabine SPUMANS
Rhaeticus 30° A Messier 60° E
Flammarion Delambre Hypatia MARE
Herschel Hipparchus Torricelli FECUNDITATIS
Ptolemaeus Capella Gutenberg Langrenus
Albategnius Theophilus Isidorus Goclenius Kapteyn
Alphonsus Abulfeda Cyrillus Mädler
Argelander Almanon Tacitus MARE Colombo Vendelinus
Airy Catharina NECTARIS
Arzachel Abenezra Geber Beaumont
Thebit La Caille Azophi Fracastorius Santbech
Purbach Playfair Sacrobosco Petavius
Blanchinus Humboldt
Werner Apianus Piccolomini Snellius
Regiomontanus Pontanus 30°
Walther Aliacensis Zagut Reichenbach
Nonius Rabbi Levi Stevinus
Orontius Büsching Riccius Furnerius
Stöfler Buch Metius Rheita
Nasireddin Maurolycus Fabricius
Saussure Licetus Barocius
Maginus Cuvier Baco Pitiscus
Jacobi Vlacq MARE AUSTRALE
Hommel
Curtius Manzinus 60°
Moretus

SOUTH

31

Moon phases

The diagram on this page shows how the different appearance of the Moon occurs. The changes in its apparent shape are purely related to the position of the Moon in its orbit around the Earth. They are not, as some people mistakenly believe, caused by the shadow of the Earth on the Moon. (The only time that happens is during a rare lunar eclipse, as described on pages 10–11. Such events occur only at Full Moon, when the Sun and Moon are on opposite sides of the Earth.) The diagram shows that New Moon is when the Moon lies between the Sun and Earth, and Full Moon when the Earth is between the Sun and the Moon. The age of the Moon (in days) is reckoned from the time of New Moon. After New Moon we have a waxing crescent until First Quarter, after which the Moon is described as waxing gibbous. After Full Moon we have a waning gibbous Moon until Last Quarter. Following that, we have a waning crescent until the next New Moon, when the sequence repeats.

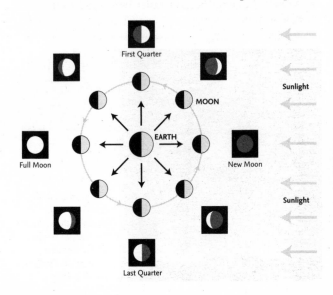

First Quarter

MOON

Sunlight

EARTH

Full Moon

New Moon

Sunlight

Last Quarter

The Circumpolar Constellations

The northern circumpolar constellations

Learning the patterns of the stars, the constellations and asterisms is not particularly difficult. You need to start by identifying the various constellations that are circumpolar where you live. These are always above the horizon, so you can generally start at any time of the year. The charts on pages 34 and 36 show the northern and southern circumpolar constellations, respectively. The fine, dashed lines indicate the areas that are circumpolar at different latitudes.

The key constellation when learning the pattern of stars in the northern sky is **Ursa Major**, in particular the seven stars forming the asterism known to many as the **'Plough'** or to people in North America as the **'Big Dipper'**. As the chart shows, this is just circumpolar for anyone at latitude 40°N, except for **Alkaid** (η Ursae Majoris), the last star in the 'tail'. Even so, the asterism of the Plough is low on the northern horizon between September and November, so it will be much easier to make out at other times of the year.

The position of the Big Dipper (the Plough), throughout the year, in relation to the northern horizon and Polaris, the Pole Star.

33

The northern circumpolar constellations.

The two stars **Dubhe** and **Merak** (α and β Ursae Majoris) are known as the 'Pointers', because they indicate the position of **Polaris**, the Pole Star (α Ursae Minoris), at about a distance of five times their separation. Following this line takes you to the constellation of **Ursa Minor**, the 'Little Bear' or 'Little Dipper', where Polaris is at the end of the 'tail' or 'handle'.

On the far side of the Pole is the constellation of **Cassiopeia**, which is highly distinctive, with its five main stars forming the letter 'M' or 'W', depending on its orientation. Cassiopeia is circumpolar for observers at latitude 40°N or closer to the North Pole, although at times it is near the northern horizon and more difficult to see. (But at such times Ursa Major is clearly visible.) To find Cassiopeia from Ursa Major, start at **Alioth** (ε Ursae Majoris) and extend a line from that star to Polaris and beyond. It points to the central star of the five.

Moving anticlockwise from Cassiopeia, we come to **Cepheus**, which has been likened to the gable end of a house, with its base in the Milky Way. The line from the Pointers to Polaris, if extended points to **Errai** (γ Cephei), at the top of the 'gable'. Continuing in the same direction, we come to **Draco**, which wraps around Ursa Minor. The quadrilateral of stars that forms the 'head' of Draco is just circumpolar for observers at 40°N latitude, although it is brushing the horizon in January. On the opposite side of the sky to the head of Draco, the whole of the faint constellation of **Camelopardalis** is visible.

For observers slightly farther north, say at 50°N, additional constellations become circumpolar. The most important of these are **Perseus**, not far from Cassiopeia, and most of which is visible and, farther round, the northern portion of **Auriga**, with bright **Capella** (α Aurigae). On the other side of the sky is **Deneb**, the brightest star in **Cygnus**, although it is often close to the horizon, especially during the early night during the winter months. **Vega** (α Lyrae) another of the three stars that form the Summer Triangle is even farther south, often brushing the northern horizon, and only truly circumpolar and clearly seen at any time of the year for observers at 60°N.

Such far northern observers will also find that **Castor** (α Geminorum) is actually circumpolar, although at times it is extremely low on the horizon. The other bright star in **Gemini**, **Pollux** (β Geminorum) is slightly farther south and cannot really be considered circumpolar.

The southern circumpolar constellations.

Eta Carinae (η Carinae) is one of the most massive and luminous stars known. It is estimated to have a mass between 120 and 150 times that of the Sun, and be between four and five million times as luminous.

The southern circumpolar constellations

Just as Ursa Major is the key constellation in the northern sky, so is **Crux** (the Southern Cross) an easily recognized feature of the southern circumpolar sky, although at times it may be brushing the horizon for observers at 30°S – roughly the latitude of Sydney in Australia. This is particularly true in the southern spring. Northeners, new to the southern sky, sometimes mistake the '**False Cross**', which consists of two stars from each of the constellations of **Vela** and **Carina** for the true Southern Cross. Crux itself is accompanied by the clearly visible dark cloud of the 'Coalsack' and also the two brightest stars in **Centaurus: Rigil Kentaurus** and **Hadar** (α and β Centauri, respectively). Together, the four stars of Crux and the two from Centaurus act as principal guides to the southern constellations.

Unfortunately, unlike the situation in the north, there is no star conveniently located at the South Celestial Pole (SCP), which lies in a relatively empty region of sky in the faint, triangular constellation of **Octans**. Octans itself is perhaps best found by using the stars of **Pavo** as guides. A line from **Peacock** (α Pavonis) through the second brightest star (β) in that constellation, if

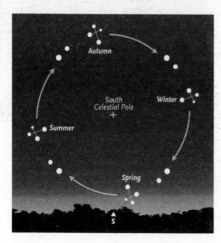

The position of Crux, the Southern Cross, throughout the year, in relation to the southern horizon. It also shows the position of the two brightest stars in Centaurus.

extended by about the same amount as the distance between the two stars, points to the 'base' of the triangle of Octans.

The main 'upright' of Crux, if extended and curving slightly to the right, does point in the approximate direction of the Pole, passing through **Musca** and the tip of **Chamaeleon**. However, a better way is to start at Hadar (the star in the bright pair that is closest to Crux), turn at right-angles at Rigil Kentaurus, and following an imaginary line through the brightest star in the small constellation of **Circinus** and then right across the sky, brushing past the outlying star of **Apus**, and the star (δ) at the apex of Octans itself.

Centaurus is a straggling constellation, with many stars well north of Rigil Kentaurus and Hadar, and some fainter ones that partially enclose Crux. Starting at Crux, and moving clockwise (in the same direction as the sky rotates), we come to the stars of **Carina** and **Vela**, both originally part of the large, now obsolete constellation of Argo Navis. Past the False Cross, we come to **Canopus** (α Carinae), the second brightest star in the sky, which is just circumpolar for observers at 40°S, although occasionally, especially in June, very low over the northern horizon. Lying between Canopus and the SCP is the **Large Magellanic Cloud** (LMC), a satellite galaxy to our own. It lies across the boundaries of the constellations of **Dorado** and **Mensa**.

Continuing round from Canopus we pass the constellations of Dorado, the small constellation of **Reticulum** and the

undistinguished constellation of **Horologium**, beyond which
is **Achernar** (α Eridani) the brightest star in the long, winding
constellation of **Eridanus**, which actually starts far to the north,
close to **Rigel** in **Orion**. Between Achernar and the SCP lies the
triangular constellation of **Hydrus**, next to the constellation of
Tucana which contains the **Small Magellanic Cloud** (SMC).

For observers farther south (at say, 50°S) there are the
constellations of **Phoenix**, followed by the roughly cross-shaped
constellation of **Grus** and the rather faint **Indus**. For observers
at 40°S, the whole of the constellation of Pavo is visible,
including its brightest star, Peacock. Farther round there is the
constellation of **Ara** and, for observers farther north at 30°S,
Triangulum Australe is fully visible.

Of the 88 constellations, the largest is **Hydra** with an area of 1303
square degrees. It covers more than 7 hours of right ascension (105°
in the sky). It is more than 19 times as large as the smallest, **Crux**,
which has an area of just 68 square degrees. **Equuleus** (87th), is not
much larger with an area of 72 square degrees. **Sagitta** is the 86th,
area 80 square degrees, and **Circinus**, 85th, with an area of 93 square
degrees. None of these four small constellation includes any objects
of particular interest.

The Monthly Maps

How to use the monthly maps

The charts in this book are designed to be used more-or-less anywhere in the world. They are not suitable to be used at very high northern or southern latitudes (beyond 60°N or 60°S). That is slightly less than the latitudes of the Arctic and Antarctic Circles, beyond which there are approximately six months of daylight, followed by six months of darkness. The design may seem a little complicated, but these diagrams should make their usage clear. The main charts are given in pairs, one pair for each month: Looking North and Looking South.

Obviously, the region of the sky that is visible at any time entirely depends on one's location on Earth. You should imagine a rectangular 'window', 90° high, that includes the sky from the horizon to the zenith. Think of moving this 'window' north or south over the charts, depending on your actual latitude. The base will be at your actual latitude on Earth. The other edge will be at your zenith. (You could make an actual 'window', by drawing a rectangle on a thin sheet of plastic, with the horizon and zenith lines 90° apart, and then use this on the charts.) The diagram on the next page shows this 'window' for the latitude of 50°N.

The scales on the right-hand and left-hand margins indicate the northern (or southern) horizon, for looking north (or south), respectively. The two diagrams on page 42 are drawn to indicate the horizon for the latitude of 40°N (the latitude of Philadelphia in the United States or Madrid in Spain); the second pair on page 43 show what would be visible 'looking north' and 'looking south' from latitude 30°S (the latitude of Durban in South Africa). If you are looking north (or south), once you get to the zenith, you can switch to the other chart, showing the view from the southern (or northern) horizon to the zenith.

To help you choose the correct latitude, there is a world map on pages 44–45.

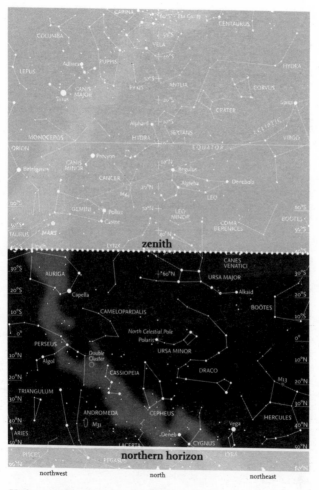

Horizon window, from the northern horizon (solid line at the bottom) to the zenith (the dotted line) for the latitude of 50°N.

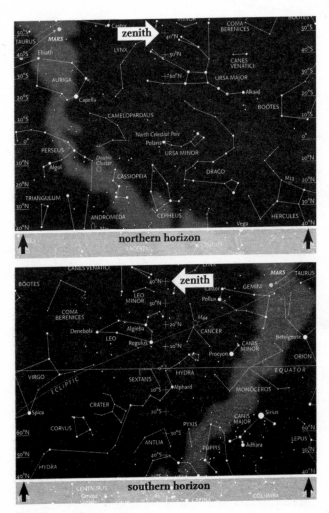

Horizons for latitude 40°N.

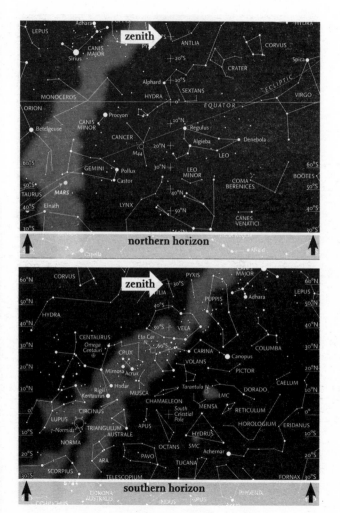

Horizons for latitude 30°S.

World Map

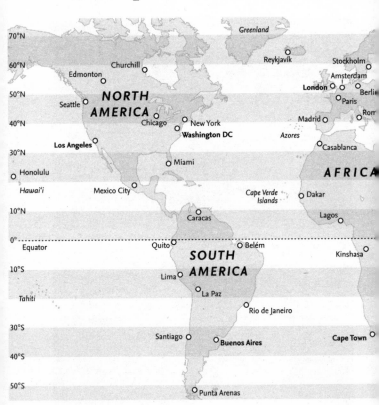

January

January – Introduction

The Earth

The Earth reaches perihelion (the closest point to the Sun in its yearly orbit) on January 3 at 03:39 Universal Time, when its distance from the Sun is 0.983306994 AU (147,100,997 km).

For northern-hemisphere observers, the **Quadrantid** meteor shower reaches maximum on the night of January 3–4, when the Moon is around Last Quarter, so conditions are reasonably favourable.'

It is one of the strongest showers of the year, with rates comparable with the Perseids (in August) and the Geminids (in December). However, peak activity is very short and easily missed, although the stream occasionally produces bright fireballs.

The Planets

Mercury comes to western elongation on January 12. *Venus* is bright (at mag. -4.0), not far from Mercury in the morning sky. *Mars* (which is mag. 1.3 to 1.4) is in the same area, but lower in the sky. *Jupiter* (mag. -2.4 to -2.5) is in southern *Aries* and *Saturn* (at mag. 1.0) is in *Aquarius*. *Uranus* (mag. 5.7) is in *Aries* and *Neptune* (very faint at mag. 7.9) is in *Pisces*.

Quadrans Muralis

The Quadrantid meteor shower perpetuates the name of an obsolete constellation, *Quadrans Muralis*, the Mural Quadrant. The instrument consisted of a quadrant mounted on a north-south wall, and used for measuring stellar positions. The area now forms part of the constellations of Boötes and Draco (see facing page).

J

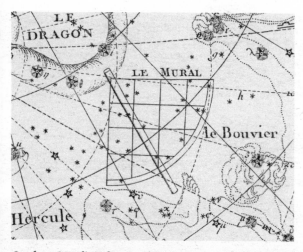

*Quadrans Muralis is shown under its original name of 'Le Mural'
in this chart from Jean Fortin's Atlas Céleste.*

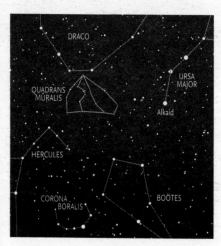

The area of the
former constellation
of Quadrans
Muralis, now part
of the constellations
of Boötes and
Draco.

Sunrise and sunset

City	Date	Sunrise	Sunset
Buenos Aires, Argentina			
	Jan. 01	08:44	23:10
	Jan. 31	09:12	23:01
Cape Town, South Africa			
	Jan. 01	03:38	18:00
	Jan. 31	04:06	17:52
London, UK			
	Jan. 01	08:07	16:02
	Jan. 31	07:42	16:47
Los Angeles, USA			
	Jan. 01	14:59	00:54
	Jan. 31	14:51	01:21
Nairobi, Kenya			
	Jan. 01	03:30	15:42
	Jan. 31	03:41	15:51
Sydney, Australia			
	Jan. 01	18:48	09:09
	Jan. 31	19:16	09:01
Tokyo, Japan			
	Jan. 01	21:51	07:38
	Jan. 31	21:42	08:06
Washington, DC, USA			
	Jan. 01	12:27	21:57
	Jan. 31	12:16	22:28
Wellington, New Zealand			
	Jan. 01	16:52	07:57
	Jan. 31	17:26	07:43

NB: the times given are in Universal Time (UT)

The Moon's phases and ages

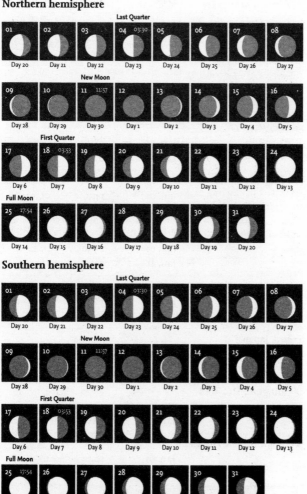

The Moon

The Moon in January
At Last Quarter on January 4, the Moon is 2° north of **Spica** in **Virgo**. It is 0.8° north of **Antares** on January 8, and then, a few hours later, passes 5.7° south of **Venus**. It is south of **Mercury** the next day, and 4.2° south of **Mars** on January 10. On January 14, the Moon is 2.1° south of **Saturn**, and close (1.0° south) to the much fainter **Neptune** the next day. On January 18, at First Quarter, the Moon is 1.8° north of **Jupiter** and 3.0° north of **Uranus** the following day (January 19). It passes 9.5° north of **Aldebaran** on January 21. By January 24, a day before Full Moon, it is 1.7° south of **Pollux**. On January 27 the Moon passes 3.9° north of **Regulus**.

Wolf Moon
Because the howling of wolves is often heard in North America in winter, the Full Moon in January is often known as the 'Wolf Moon'. The name may originally stem from the Old-World, Anglo-Saxon lunar calendar. Other names for this Full Moon include: Moon After Yule, Old Moon, Ice Moon, and Snow Moon. Among the Algonquin tribes, the name was 'squochee kesos', meaning 'the Sun has not strength to thaw'. The name 'Wolf Moon' was occasionally applied to the Full Moon in December.

In this photograph, the narrow lunar crescent (about two days old) has been over-exposed to show the Earthshine illuminating on the other portion of the Moon, where the dark maria are faintly visible.

Names of the Full Moon in North America

Many different cultures had specific names for the Full Moon, depending on the time of year. Even in a single culture, the actual names often varied between different tribes, so there may be more than one name used for a particular Full Moon. The interval between successive Full Moons (or between any other specific phases of the Moon) is known as the synodic month, and is, on average, 29.53 days, so the names have come to be associated with modern calendar months. One of the most commonly known sets of names is that used by the various tribes in North America. These are:

- January: 'Wolf Moon'
- February: 'Snow Moon', 'Hunger Moon', 'Storm Moon'
- March: 'Worm Moon', 'Crow Moon', 'Sap Moon', 'Lenten Moon'
- April: 'Seed Moon', 'Pink Moon', 'Sprouting Grass Moon', 'Egg Moon', 'Fish Moon'
- May: 'Milk Moon', 'Flower Moon', 'Corn Planting Moon'
- June: 'Mead Moon', 'Strawberry Moon', 'Rose Moon'
- July: 'Hay Moon', 'Buck Moon', 'Elk Moon', 'Thunder Moon'
- August: 'Corn Moon', 'Sturgeon Moon', 'Red Moon', 'Green Corn Moon', 'Grain Moon'
- September: 'Harvest Moon', 'Full Corn Moon'
- October: 'Hunter's Moon', 'Blood Moon'/'Sanguine Moon'
- November: 'Beaver Moon', 'Frost Moon'
- December: 'Oak Moon', 'Cold Moon', 'Long Night Moon'

The only two names commonly used in Europe were 'Harvest Moon' and 'Hunter's Moon'. On rare occasions, particularly in religious contexts, the term 'Lenten Moon' was used for the Full Moon in March. The other terms, which originated in North America, have been adopted increasingly by the media in recent years.

Calendar for January

Dec.28–Jan.12		Quadrantid meteor shower
01	15:28	Moon at apogee = 404,909 km
03	03:39	Earth at perihelion (147,100,997 km = 0.9833 AU)
03–04		Quadrantid meteor shower maximum
04	03:30	Last Quarter
04	23:48	Spica 2.0°S of the Moon
08	14:59	Antares 0.8°S of the Moon
08	20:10	Venus (mag. -4.0) 5.7°N of the Moon
09	18:47	Mercury (mag. -0.2) 6.6°N of the Moon
10	08:31	Mars (mag. 1.4) 4.2°N of the Moon
11	11:57	New Moon
12	14:37	Mercury at greatest elongation (23.5°W, mag. -0.3)
13	10:36	Moon at perigee = 362,266 km
14	09:33	Saturn (mag. 1.0) 2.1°N of the Moon
15	20:25	Neptune (mag. 7.9) 1.0°N of the Moon
18	03:53	First Quarter
18	20:42	Jupiter (mag. -2.5) 1.8°S of the Moon
19	19:39	Uranus (mag. 5.7) 3.0°S of the Moon
21	11:06	Aldebaran 9.5°S of the Moon
24	19:39	Pollux 1.7°N of the Moon
25	17:54	Full Moon
27	16:00 *	Mars (mag. 1.3) 0.2°S of Mercury (mag. -0.2)
27	16:59	Regulus 3.9°S of the Moon
29	08:14	Moon at apogee = 405,777 km
31–Feb.20		α-Centaurid meteor shower

These objects are close together for an extended period around this time.

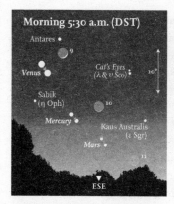

January 9–11 • *The narrow crescent Moon passes Antares, Venus, Mercury and Mars (as seen from Sydney).*

January 14 • *The Moon with Saturn, side by side in the western sky (as seen from Sydney).*

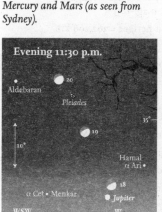

January 18–20 • *The Moon passes Jupiter, and the Pleiades and Aldebaran. Menkar and Hamal are nearby (as seen from London).*

January 24 • *The almost Full Moon lines up with Pollux and Castor. Procyon is closer to the horizon (as seen from London).*

January – Looking North

For most northern observers, all the important northern circumpolar constellations (see pages 33–35) of **Ursa Major**, **Ursa Minor** (with **Polaris**, the Pole Star), **Cassiopeia**, **Draco** and **Cepheus** will be visible. Polaris is, of course, the star about which the sky appears to rotate, even though it is not precisely at the North Celestial Pole. Ursa Major stands more-or-less vertically above the horizon in the northeast. Opposite it in the northwest of the sky is the 'W' of Cassiopeia. (Looking more like an 'M' at this time of the year.) For most observers, **Capella** (α Aurigae) is high overhead, but only those at high latitudes will find it easy to see the quadrilateral of stars that marks the head of Draco, brilliant **Deneb** (α Cygni) or the even brighter **Vega** (α Lyrae), yet farther south. Deneb and **Eltanin** (γ Draconis), the brightest star in the 'Head' of Draco, are skimming the horizon for observers at 40°N.

On 2 January 2019, the Chinese lunar probe **Chang-e 4** became the first object to achieve a landing on the far side of the Moon. It landed near the South Pole in the Von Karman crater, which is itself in the South Pole-Aitken basin.

For observers between about 30 and 50°N, the constellation of **Auriga** is near the zenith (and thus difficult to observe). This important constellation contains bright **Capella** (α Aurigae) and farther to its west lies the constellation of **Perseus**, with **Algol**, the famous variable star.

The constellation of Auriga, with brilliant Capella which, although appearing as a single star, is actually a quadruple system, consisting of a pair of yellow giant stars, gravitationally bound to a more distant pair of red dwarfs. Elnath, near the bottom, actually belongs to the constellation of Taurus.

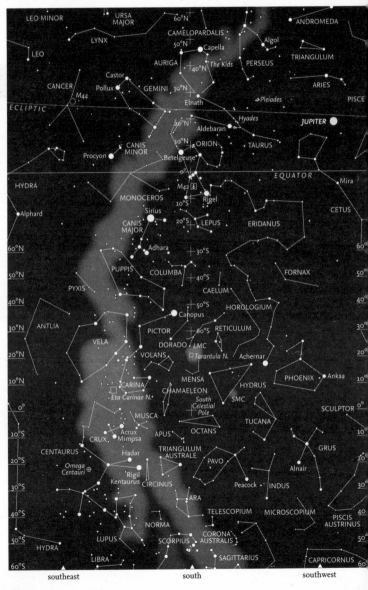

LEO MINOR
URSA MAJOR
60°N
CAMELOPARDALIS
ANDROMEDA
LYNX
Algol
Capella
50°N
AURIGA
40°N
The Kids
PERSEUS
TRIANGULUM
LEO
Castor
30°N
Pollux
GEMINI
ARIES
CANCER
M44
Elnath
Pleiades
PISCE
ECLIPTIC
20°N
JUPITER
Aldebaran
Hyades
10°N
ORION
TAURUS
Procyon
CANIS MINOR
Betelgeuse
EQUATOR
Mira
HYDRA
M42
MONOCEROS
Rigel
CETUS
Alphard
10°S
Sirius
LEPUS
ERIDANUS
CANIS MAJOR
20°S
60°N
Adhara
30°S
50°N
PUPPIS
COLUMBA
FORNAX
50°
40°S
PYXIS
CAELUM
40°
50°S
HOROLOGIUM
30°N
Canopus
60°S
RETICULUM
30°
ANTLIA
PICTOR
DORADO LMC
20°N
VELA
VOLANS
Achernar
10°
10°N
CARINA
MENSA
HYDRUS
PHOENIX
Ankaa
Eta Carinae N.
CHAMAELEON
Tarantula N.
SMC
SCULPTOR
0°
South Celestial Pole
MUSCA
TUCANA
10°S
Acrux
APUS
Mimosa
OCTANS
GRUS
CENTAURUS
CRUX
TRIANGULUM AUSTRALE
20°S
Hadar
PAVO
Alnair
Omega Centauri
Rigil Kentaurus
CIRCINUS
30°S
Peacock
INDUS
ARA
40°S
TELESCOPIUM
MICROSCOPIUM
PISCIS AUSTRINUS
NORMA
50°S
HYDRA
CORONA AUSTRALIS
LUPUS
SCORPIUS
LIBRA
SAGITTARIUS
CAPRICORNUS
60°S

southeast   south   southwest

58

January – Looking South

The southern sky is dominated by **Orion**, visible from nearly everywhere in the world and prominent during the northern winter months. For observers near the equator it is, of course, high above near the zenith. Orion is highly distinctive, with a line of three stars that form the 'Belt'. To most observers, the bright star **Betelgeuse** (α Orionis), shows a reddish tinge, in contrast to the brilliant bluish-white **Rigel** (β Orionis). The three stars of the belt lie directly south of the celestial equator. A vertical line of three 'stars' forms the 'Sword' that hangs south of the Belt. With good viewing, the central 'star' appears as a hazy spot, even to the naked eye, and is actually the **Orion Nebula** (M42). Binoculars reveal the four stars of the Trapezium, which illuminate the nebula.

Orion's Belt points up to the northwest towards **Taurus** (the Bull) and orange-tinted **Aldebaran** (α Tauri). Close to Aldebaran, there is a conspicuous 'V' of stars, called the **Hyades** cluster. (Despite appearances, Aldebaran is not part of the cluster.) Farther along, the same line from Orion passes below a bright cluster of stars, the **Pleiades**, or Seven Sisters. Even the smallest pair of binoculars reveals this as a beautiful group of bluish-white stars. The two most conspicuous of the other stars in Taurus lie directly north of Orion, and form an elongated triangle with Aldebaran. The northernmost, **Elnath** (β Tauri), was once considered to be part of the constellation of **Auriga**.

Slightly to the west of **Capella** lies a small triangle of fainter stars, known as '**The Kids**'. (Ancient mythological representations of Auriga show him carrying two young goats.) Together with Elnath, the body of Auriga forms a large pentagon on the sky, with The Kids lying on the western side (see page 57).

Running south from Orion is the long constellation of **Eridanus** (the River), which begins near Rigel in Orion and runs far south to end at **Achernar** (α Eridani). To the south of Orion is the constellation of **Canis Major** and several other constellations, including the oddly shaped **Carina**. The line of Orion's Belt also points southeast in the general direction of **Sirius** (α Canis Majoris), the brightest star. Almost due south of Sirius lies **Canopus** (α Carinae), the second brightest star in the sky.

Simon Marius

Simon Marius was born on 10 January 1573. (He was born Simon Mayr, and 'Marius' is a latinized version of his surname.) He studied in Heilsbronn and became an astronomer. In 1601 he went to Prague to study under Tycho Brahe. He moved to Padua in December 1601.

Marius had a famous dispute with Galileo Galilei over the discovery of the Jovian satellites. This remained an unsettled question for some 300 years, until a Dutch investigation in 1903 established that Marius did indeed discover the moons. His notes show that he observed the satellites on 29 December 1609 (OS), one day before Galileo described the satellites in a letter of 8 January 1610 (NS). Part of the problem arose from the fact that Marius used the older Julian calendar (OS) and Galileo the new Gregorian calendar (NS).

A lunar crater in Oceanus Procellarum is named after Marius (see facing page). It is about 41 km in diameter and some 1.7 km deep. To the west and north of the crater there are a large number of domes, though to have been created by volcanic action, known as the Marius Hills (see page 62).

Simon Marius, as depicted in an engraving in his book Mundus Iovialis, *published in 1614.*

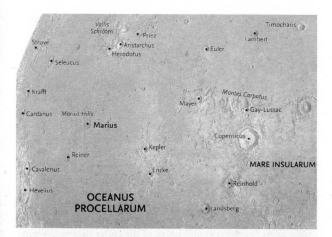

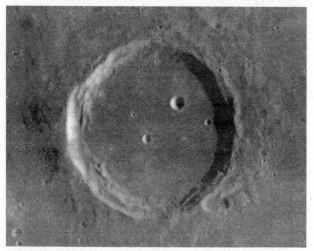

The lunar crater Marius (diameter 41 km), as imaged by the Lunar Orbiter 4 *spacecraft.*

The Marius Hills

The Marius Hills are an unusual feature of the Moon. They lie in the Oceanus Procellarum, west and north of the crater Marius. They appear to be volcanic lava domes, and are the largest concentration of volcanic features on the Moon. They vary in height from 200 to 500 metres, and spread over an area that is approximately 100 km across. The lava from which they formed appears to have been far more viscous than the lava that flooded the mare basins. The area also contains rilles, wrinkle ridges and various channels that were presumably formed at the same time.

On 16 February 2019, the Lunar Reconnaissance Orbiter obtained a vertical image of the Marius Hills in Oceanus Procellarum. The probe has photographed a pit, which might be a 'skylight', where part of the roof of an underground lava tube has collapsed, after the lava has drained away. Such pits might form suitable locations for protecting future manned habitations from cosmic radiation and temperature extremes.

An oblique view of the Marius Hills as obtained by the Lunar Orbiter 2 *spaceprobe.*

February

February – Introduction

There are just two, relatively minor, meteor showers in February 2024, which is a quiet month for astronomers with few notable events. The two showers are both southern ones, and neither is visible to northern observers. The most significant shower is the *α-Centaurids*, which actually begins on 31 January 2024. This shower has two separate branches, with radiants lying near the brightest stars in Centaurus (**Rigil Kentaurus** and **Hadar**, or α and β Cen, respectively). These two streams are thus known as the *α-Centaurids* and the *β-Centaurids*. Both branches of this shower reach a low maximum, with an hourly rate of about 5–6 meteors per hour, on February 8. That day the Moon is at Day 28 of the lunation, just before New Moon, so observing conditions are very favourable.

The second southern shower, the *γ-Normids*, begins its activity around 25 February 2024 and continues into March, reaching a weak (but very sharp) maximum on March 14–15. Unfortunately, its meteors are difficult to differentiate from sporadics, so are likely to be identified by dedicated meteor observers only. Observing conditions at maximum are reasonably favourable in 2024, because the Moon is a waxing crescent (Days 4 and 5 of the lunation).

The planets

Mercury is 3.2° north of the Moon on February 8, one day before New Moon. **Venus** remains bright (mag. -3.9) in the morning sky, but becomes lower towards the horizon. **Mars** is too close to the Sun to be readily visible. **Jupiter** is moving slowly in Aries at mag. -2.3 to -2.2. **Saturn** (mag. 1.0) remains in **Aquarius**. **Uranus** (mag. 5.7 to 5.8) is in **Aries** and **Neptune** (mag. 7.9 to 8.0) in Pisces.

Why is February short?

Why does February have such an odd number of days, and why do we tinker with it every four years? The answer is suprisingly complicated, and involves the ancient Roman lunar calendar, Roman emperors, including Julius Caesar, the Roman Senate, the priests, and the way in which politicians messed about with the calendar, and how we have avoided even greater confusion.

Occultations

Although there are as many as fourteen occultations of Antares in 2024 and nine of Spica, none are readily visible from land, visible (if at all) from the oceans. There are no occultations in 2024 of the other bright stars (Aldebaran, Pollux and Regulus) that are near the ecliptic.

F

February 2021 was a busy month at Mars. On February 9, a spaceprobe launched by the United Arab Emirates, named *Hope*, arrived to carry out atmospheric studies. The next day, the Chinese *Tianwen-1* mission arrived. This mission consists of an orbiter, a lander and a rover. The lander and rover landed on 14 May 2021 and the *Zhurong* lander was deployed on 22 May 2021. On 18 February 2021, the NASA Perseverance rover, with its helicopter *Ingenuity* was successfully landed in Jezero crater.

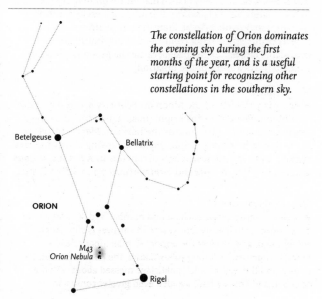

The constellation of Orion dominates the evening sky during the first months of the year, and is a useful starting point for recognizing other constellations in the southern sky.

The names and lengths of the months

Several modern names for months derive from the original Roman 'calendar of Romulus', which contained just ten months. The winter was considered to be without Full Moons and thus months. The calendar dates back to around 750 BCE. The legendary king Romulus was supposed to have started naming the months, beginning at the spring equinox (in March). So we have:

- March from **Martius** (Mars, the god of war)
- April from **Aprilis** (the reason for this one is uncertain, and possibly related to the raising of hogs)
- May from **Maius** (a local Italian goddess)
- June from **Junius** (the queen of the Latin gods).

After Junius, he simply counted the months beginning **Quintilis**, **Sextilis** (our July and August). This gave:

- September from **septem** (seven)
- October from **octo** (eight)
- November from **novem** (nine)
- December from **decem** (ten)

About 713 BCE, two more, January (**Januarius** from Janus, the Roman god of beginnings) and February (**Februarius**, from Februa, the Roman ritual of purification at the end of the year) were added.

Originally, the Romans used a lunar calendar, beginning at the spring equinox in March. To keep the lunar calendar in step with the year, what are called intercalary months were sometimes inserted.

The calendar was initially under the control of priests and they fiddled things to their advantage. The calendar was used to determine when people could carry out business, and when ceremonies could occur. The priests added intercalary months whenever they thought fit, for example when more taxes were required. There was utter chaos and the calendar never agreed with the solar year.

Eventually, Julius Caesar instigated the Julian calendar reform. Months alternated between 30 and 31 days. This was slightly too long, giving 366 days, so one day was removed from the last month of the year, February, to give it 29 days. An additional day was returned to February every four years, to keep things in step with the Sun. One month, Quintilis, was named Julius (the month of Caesar's birth).

But then the priests messed things up again. They started counting leap years every three years. The error was corrected by the emperor Augustus and by 8 CE the matter had been solved and the months and the Sun were in agreement. But then the Senate decided to rename one month in honour of Augustus – so the month of Sextilis became our August. Unfortunately, under Caesar's scheme that month had just 30 days, whereas Caesar's (our July) had 31 days. Obviously Augustus had to have the same number of days, so they pinched one from poor February, leaving it with 28 days, except in leap years. At the same time, to avoid having three months with 31 days in succession they also tinkered with the lengths of the months after August, which is why September and November now have 30 days and October and December 31.

Two later emperors tried to change things again. Domitian (emperor between 81 and 96 CE) wanted to rename September, the month when he became emperor, to *Gemanicus* to commemorate his victory over the German tribes, and then rename October (when he was born) to *Domitianus*. However the Roman Senators did not approve of Domitian, who had drastically reduced their powers, so they retained the changes initiated by Augustus. Even worse might yet have ensued. A later emperor was Commodus (emperor 176–192 CE), the son of emperor Marcus Aurelius. Towards the end of his reign, he became a megalomaniac, adopting various names. After a disastrous fire devastated Rome in 191 CE, early in 192 Commodus declared himself the new Romulus, and ritually re-founded Rome, renaming the city *Colonia Lucia Annia Commodiana*. He then tried to rename all the months of the year to correspond exactly with his own (now twelve) names: *Lucius*, *Aelius*, *Aurelius*, *Commodus*, *Augustus*, *Herculeus*, *Romanus*, *Exsuperatorius*, *Amazonius*, *Invictus*, *Felix* and *Pius*.

Sunrise and sunset

City	Date	Sunrise	Sunset
Buenos Aires, Argentina			
	Feb. 01	09:13	23:00
	Feb. 28	09:40	22:31
Cape Town, South Africa			
	Feb. 01	04:07	17:52
	Feb. 28	04:33	17:23
London, UK			
	Feb. 01	07:40	16:49
	Feb. 28	06:48	17:40
Los Angeles, USA			
	Feb. 01	14:51	01:22
	Feb. 28	14:22	01:48
Nairobi, Kenya			
	Feb. 01	03:41	15:51
	Feb. 28	03:41	15:49
Sydney, Australia			
	Feb. 01	19:17	09:00
	Feb. 28	19:43	08:33
Tokyo, Japan			
	Feb. 01	21:41	08:07
	Feb. 28	21:11	08:35
Washington, DC, USA			
	Feb. 01	12:15	22:29
	Feb. 28	11:41	23:01
Wellington, New Zealand			
	Feb. 01	17:27	07:42
	Feb. 28	18:02	07:05

NB: *the times given are in Universal Time (UT)*

The Moon's phases and ages

Northern hemisphere

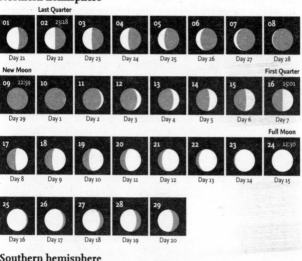

Southern hemisphere

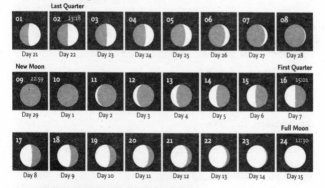

The Moon

The Moon in February

On February 1, just before Last Quarter, the Moon is 1.7° north of **Spica**. By February 5 it is 0.6° north of **Antares** in **Scorpius** and on February 7 it is 5.7°S of brilliant **Venus**. It passes 4.2° south of **Mars** the next day and 3.2° south of **Mercury** later that same day (February 8). Two days after New Moon (by February 11) the Moon is 1.8° south of **Saturn** and the next day is even closer to (0.7° south) the much fainter **Neptune**. Three days later (by February 15) the Moon is 3.2° north of **Jupiter** and the next day the same distance north of **Uranus**. On February 17, one day after First Quarter, the Moon is 9.8°N of **Aldebaran** in **Taurus**. By February 21, it is 1.6°S of **Pollux**. Three days after Full Moon, on February 27, the Moon is 3.9° north of **Regulus**, passing between it and **Algieba**.

Snow Moon

In the northern hemisphere, February is often the coldest month, and most countries on both sides of the Atlantic see significant falls of snow. The Full Moon of February is thus often called the 'Snow Moon', although just occasionally that name has been applied to the Full Moon in January. Some North American tribes named it the 'Hunger Moon' because of the scarcity of food sources during the depths of winter, while other names are 'Storm Moon' and 'Chaste Moon', although the last name is more commonly applied to the Full Moon in March. To the Arapaho of the Great Plains, the Full Moon was called the Moon 'when snow blows like grain in the wind'.

Black Moon

Because it never has more than 29 days, and the synodic month (between any pair of phases, such as from New Moon to New Moon) is slightly more than half a day longer, sometimes there is no Full Moon in February. This occurs about every 19 years. This is one of the definitions of the term **'Black Moon'**, although the same term is sometimes applied when there is no New Moon in a particular calendar month.

The Winchcombe meteorite

At 21:54 on the evening of 28 February 2021, a brilliant fireball (a bolide) was observed over Gloucestershire. Observations allowed the orbit of the parent body to be determined, showing that the body's original location was the outer region of the asteroid belt between Mars and Jupiter. One fragment was recovered from the driveway of a house in the market town of Winchcombe. The largest fragment from this fall weighed 152 grammes.

The Winchcombe meteorite is particularly important, not least because it was recoved extremely rapidly – before it had time to be contaminated by earthly material, but also because it belongs to a moderately rare class of stony meteorites, known as carbonaceous chondrites. (There are three main classes of meteorites, known as stones, stony-irons and irons.) Chondrules are small, spherical inclusions, which are believed to have originally formed when free-floating in space, before

The Winchcombe meteorite.

becoming incorporated into a larger body. They are estimated to be about 4550 million years old and are thus (like comets) some of the oldest objects in the Solar System.

The water contained within the Winchcombe meteorite is particularly significant. It closely matches the water found on Earth, and helps to confirm the suspicion that all water on Earth originated in meteorites rather in icy comets, where the composition of the water differs from terrestrial water.

Of the chondrites, the most important scientifically are the carbonaceous chondrites, like the Winchcombe meteorite. Apart from the chondrules, these may contain organic material (such as amino acids), water and pre-solar grains. These compounds are essential for life, and there is an opinion that life on Earth has arisen because these materials have been delivered by carbonaceous meteorites or carbonaceous minor planets that have impacted on the Earth.

The Winchcombe meteorite. Some of the chondrules may be seen as white specks within the body.

F

The Murchison meteorite, which fell on 28 September 1969 in Murchison in the state of Victoria in Australia. Some of the silicon carbide particles in this meteorite have been dated at 7000 million years old, some 2500 million years older than the age of the Solar System, 4500 million years.

Calendar for February

01	07:46	Spica 1.7°S of the Moon
02	23:18	Last Quarter
05	00:52	Antares 0.6°S of the Moon
06	06:30	Mars (mag. 1.3) 4.2°N of the Moon
06	21:59	Mercury (mag. -0.4) 3.2°N of the Moon
07	18:50	Venus (mag. -3.9) 5.4°N of the Moon
08		α-Centaurid meteor shower maximum
09	22:59	New Moon
10	18:53	Moon at perigee = 358,099 km
11	00:40	Saturn (mag. 1.0) 1.8°N of the Moon
12	06:45	Neptune (mag. 7.9) 0.7°N of the Moon
15	08:16	Jupiter (mag. -2.3) 3.2°S of the Moon
16	08:16	Uranus (mag. 5.8) 3.2°S of the Moon
16	15:01	First Quarter
17	16:41	Aldebaran 9.8°S of the Moon
21	01:33	Pollux 1.6°N of the Moon
23	23:26	Regulus 3.6°S of the Moon
24	12:30	Full Moon
25–Mar.28		γ-Normid meteor shower
25	14:59	Moon at apogee = 406,312 km
28	14:23	Spica 1.5°S of the Moon
28	14:00 *	Saturn (mag. 0.9) 0.2°N of Mercury (mag. -1.8)

These objects are close together for an extended period around this time.

February 1 • In the morning, the Moon and Spica are close together in the southwestern sky (as seen from London).

February 8–9 • The crescent Moon passes Venus, Mars and Mercury. Mars is not very bright (mag. 1.3) as seen from Sydney.

February 21 • The waxing gibbous Moon lines up with Pollux and Castor (as seen from London).

February 24 • High in the south, the Full Moon passes between Regulus and Algieba (as seen from London).

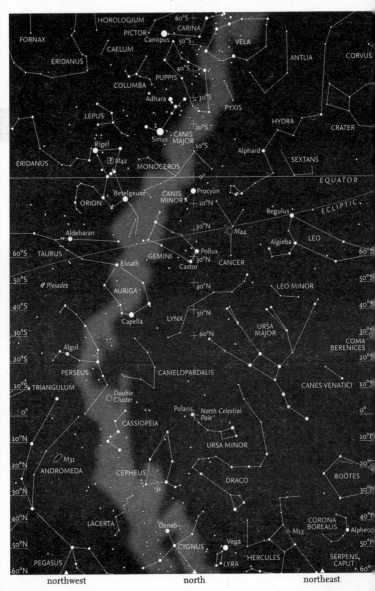

northwest — north — northeast

76

February – Looking North

The months of January and February are probably the best time for seeing the section of the Milky Way that runs in the northern and western sky from *Orion* and *Gemini* right through *Auriga*, *Perseus* and *Cassiopeia*, towards *Cygnus*, low in the north. Although not as readily visible as the denser star clouds of the summer Milky Way, on a clear night so many stars may be seen that even a distinctive constellation such as Cassiopeia, which lies across the Milky Way, is not immediately obvious.

F

The 'base' of the constellation of *Cepheus* lies on the edge of the stars of the Milky Way, but the red supergiant star *Mu* (μ) *Cephei*, called the 'Garnet Star' by William Herschel with its striking red colour remains readily visible. The groups of stars, known as the *Double Cluster* in Perseus (NGC 869 & NGC 884, often known as h and χ Persei), lying between Perseus and Cassiopeia, are well-placed for observation.

Beyond the Milky Way, Perseus and Cassiopeia, the constellation of *Andromeda* is beginning to be lost in the north-western sky.

Castor and *Pollux* in *Gemini* are near the zenith for observers at 30°N, but for most northern observers the constellation is best seen when facing south. The head of *Draco* is now higher in the sky and easier to recognize, with its long, straggling 'body' curling round *Ursa Minor* and the Pole. *Ursa Major* has begun to swing round towards the north, and Cassiopeia is now lower in the western sky.

For observers in the far north, most of *Cygnus*, with

The Japanese spacecraft *Hayabusa2* obtained the first sample of asteroid *Ryugu* on 21 February 2019.

its brightest star, *Deneb* (α Cygni), is visible in the north, and even *Lyra*, with *Vega* (α Lyrae) may be seen at times. Most of the constellation of *Hercules* is visible, together with the distinctive circlet of *Corona Borealis* to its east. Observers farther south may see Deneb and even Vega peeping over the northern horizon at times during the night, although they will often be lost (like all the fainter stars) in the inevitable extinction along the horizon.

LOOKING SOUTH — Early February 23:00 — Mid February 22:00 — Late February 21:00

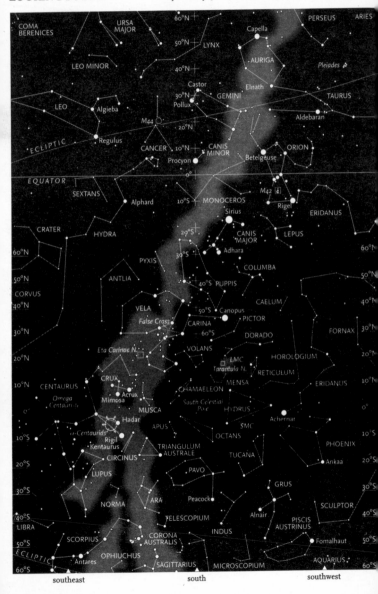

southeast   south   southwest

78

February – Looking South

Orion is now beginning to sink into the southwest, and the two brightest stars in the sky, **Sirius** and **Canopus** (α Carinae), are readily visible to observers at low northern latitudes and, of course, to those who are south of the equator. (Canopus is close to the zenith for those in the far south.)

F

South of **Carina** and the neighbouring constellation of **Vela** (both part of the original, and now obsolete, constellation of Argo Navis), lies the sprawling constellation of **Centaurus**, surrounding the distinctive constellation of **Crux**, the Southern Cross, which is on the horizon at 10°N. North of this, the 'False Cross', sometimes mistaken for the true constellation of Crux, consists of two stars from each of Vela (δ and κ Velorum) and Carina (ε and ι Carinae). The two brightest stars of Centaurus, **Rigil Kentaurus** (α Centauri) and **Hadar** (β Centauri) are slightly farther south, beyond Crux.

> **Sirius,** α Canis Majoris (α CMa), in the southern celestial hemisphere, is the brightest star in the sky at magnitude -1.44.

The **Large Magellanic Cloud** (LMC) is almost on the meridian, early in the night, to the west of Carina and the small constellation of **Volans**. It is surrounded by several small constellations: **Mensa**, **Reticulum** and **Dorado**. (Technically, it is largely within the area of Dorado.) Any optical aid (such as binoculars) will begin to show some of the remarkable structures within the LMC, including the great **Tarantula Nebula**, or *30 Doradus*, a site of active star formation.

Farther south and west lies **Achernar** (α Eridani) , the bright star at the end of **Eridanus**, the long straggling constellation that represents a river and that may now be traced all the way from where it begins near **Rigel** (β Orionis) in Orion.

Beyond Crux, and on the other side of the Milky Way, lies the rest of Centaurus. Northeast of Crux is the finest and brightest globular cluster in the sky, **Omega** (ω) **Centauri** also known as NGC 5139. It is the largest globular cluster in our Galaxy and is estimated to contain about 10 million stars. Although appearing like a star, its non-stellar nature was discovered by Edmond Halley in 1677.

March

March – Introduction

The Sun crosses the celestial equator from south to north at the vernal equinox, on Wednesday, March 20 in 2024. Day and night are then of almost equal length, although refraction in the atmosphere 'raises' the Sun above the horizon at its rising and setting, so daylight at the equinox itself is slightly longer than darkness. (The hours of daylight and darkness change most rapidly around the equinoxes in March and September.) The northern season of spring is considered to begin at the equinox and ancient calendars started the year at that date. (Until Pope Gregory revised the calendar in 1582.) It is also in March that Daylight Saving Time (DST), known in Britain as British Summer Time (BST), begins (on Saturday–Sunday, March 30–31). (In Europe, Daylight Saving Time is introduced on the same date.)

The point at which the Sun crosses the celestial equator is known as the ***First Point of Aries***, and originally lay in that constellation. Because of the phenomenon of precession, which changes the orientation of the Earth's axis in space, this point now actually lies in the constellation of ***Pisces***, close to the border with Aquarius, lying near the star λ Piscium. It is slowly moving towards Aquarius at a rate of about one degree every 70 years.

The autumnal equinox (or spring equinox for those in the southern hemisphere), when the Sun crosses the celestial equator in the opposite direction (north to south) occurs

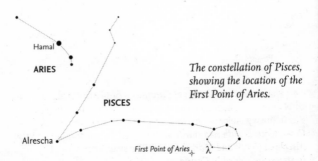

The constellation of Pisces, showing the location of the First Point of Aries.

in September (September 22 in 2024). It lies in the large constellation of **Virgo** (see page 176).

The first point of Aries is sometimes known as the Cusp of Aries, and the equinox in September as the Cusp of Libra.

The planets
Mercury is too close to the Sun to be visible this month. **Venus** is similarly placed. **Mars** is moving towards the Sun, but may be glimpsed in the morning sky early in the month. **Jupiter** (mag. -2.0) is in **Aries**, crossing into **Taurus** at the end of the month. **Saturn** (mag. 1.1 to 1.2) is moving slowly in **Aquarius**. **Uranus** (mag. 5.8) remains in **Aries** and **Neptune** (mag. 8.0 to 7.9) in **Pisces**. Minor planet **(3) Juno** (mag. 8.7) comes to opposition on March 3 (see the maps on page 96).

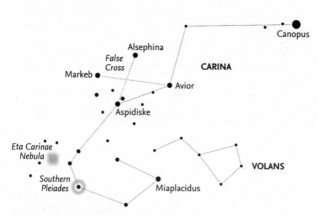

This month the constellation of Carina is well placed for observers at 20°N or more to the south. Canopus (α Car) is the second brightest star in the sky. Avior (ε Car) and Aspidiske (ι Car) form the False Cross, together with two stars that belong to the constellation of Vela: Alsephina (δ Vel) and Markeb (κ Vel). The small constellation of Volans is almost embedded in Carina.

Sunrise and sunset

City	Date	Sunrise	Sunset
Buenos Aires, Argentina			
	Mar. 01	09:41	22:29
	Mar. 31	10:06	21:49
Cape Town, South Africa			
	Mar. 01	04:34	17:22
	Mar. 31	04:58	16:42
London, UK			
	Mar. 01	06:45	17:42
	Mar. 31	05:38	18:33
Los Angeles, USA			
	Mar. 01	14:21	01:49
	Mar. 31	13:41	02:13
Nairobi, Kenya			
	Mar. 01	03:41	15:49
	Mar. 31	03:34	15:40
Sydney, Australia			
	Mar. 01	19:44	08:32
	Mar. 31	20:07	07:52
Tokyo, Japan			
	Mar. 01	21:10	08:36
	Mar. 31	20:27	09:02
Washington, DC, USA			
	Mar. 01	11:40	23:02
	Mar. 31	10:54	23:31
Wellington, New Zealand			
	Mar. 01	18:03	07:04
	Mar. 31	18:36	06:14

NB: *the times given are in Universal Time (UT)*

The Moon's phases and ages

Northern hemisphere

Last Quarter

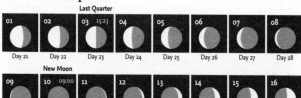

01	02	03 15:23	04	05	06	07	08
Day 21	Day 22	Day 23	Day 24	Day 25	Day 26	Day 27	Day 28

New Moon

09	10 09:00	11	12	13	14	15	16
Day 29	Day 30	Day 1	Day 2	Day 3	Day 4	Day 5	Day 6

First Quarter

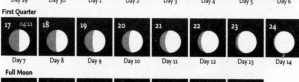

17 04:11	18	19	20	21	22	23	24
Day 7	Day 8	Day 9	Day 10	Day 11	Day 12	Day 13	Day 14

Full Moon

25 07:00	26	27	28	29	30	31
Day 15	Day 16	Day 17	Day 18	Day 19	Day 20	Day 21

Southern hemisphere

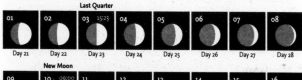
Last Quarter

01	02	03 15:23	04	05	06	07	08
Day 21	Day 22	Day 23	Day 24	Day 25	Day 26	Day 27	Day 28

New Moon

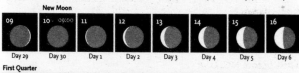

09	10 09:00	11	12	13	14	15	16
Day 29	Day 30	Day 1	Day 2	Day 3	Day 4	Day 5	Day 6

First Quarter

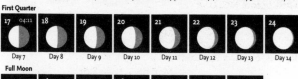

17 04:11	18	19	20	21	22	23	24
Day 7	Day 8	Day 9	Day 10	Day 11	Day 12	Day 13	Day 14

Full Moon

25 07:00	26	27	28	29	30	31
Day 15	Day 16	Day 17	Day 18	Day 19	Day 20	Day 21

M

85

The Moon

The Moon in March

On March 3, at Last Quarter, the Moon is just 0.4° north of **Antares** in **Scorpius**. On March 8 it passes 3.5° south of **Mars**. Later the same day it is 3.3°S of **Venus**. The next day (March 9) it is 1.5° south of **Saturn**. On March 10, a few hours before New Moon, it is 0.5° south of **Neptune** (which is faint at mag. 8.0) and the next day it is 1.0° south of **Mercury**. On March 14 it is 4.0° north of **Jupiter** (which is bright at mag.-2.1) and later that day 3.4° north of **Uranus** (at mag. 5.8). On March 15 it is 9.9° north of **Aldebaran** in **Taurus**. On March 19, just after First Quarter, it is 1.4° south of **Pollux**. By March 22 it is 3.6° north of **Regulus**. At Full Moon (on March 25) there is a penumbral lunar eclipse (visible from the USA) when the Moon may just touch the umbra. The next day the Moon is 1.4° north of **Spica**. On March 30 the Moon is again 0.3° north of **Antares**.

Worm Moon

One name for the last Full Moon of the winter season, which falls in March, is the 'Worm Moon'. The name derives from the fact that earthworms become active in the soil at the end of winter and are sometimes seen at the surface. Other names include 'Crow Moon', because the birds become particularly active and are avid to feed on the worms, after the lack of food during the winter months. Another name is the 'Sap Moon', which is particularly relevant in Canada, because this is the time when maple trees may be tapped for the sap (to produce maple syrup, beloved by Canadians). In Europe, the term 'Lenten Moon' was sometimes used, and this is the Old English/Anglo-Saxon name for this particular Full Moon, and is derived, of course, from the Christian period of Lent.

M

Lunar eclipse of March 25

On March 25 there will be a penumbral eclipse of the Moon, visible from a large portion of Asia and the Middle East. It will not be visible from central and western Europe. Practically the whole of the Moon will enter the Earth's penumbra, with just a small portion lying outside the shadow. Unfortunately, penumbral eclipses are so weak that, in general, nothing is detectable by the naked eye, and the diminution of light is too faint to be readily seen. When about 70 per cent or more of the Moon's diameter enters the penumbra, the darkening of the Moon's surface may just be detectable to the naked eye. In this event, however, about 95 per cent of the Moon will enter the penumbra, so some people may be able to detect the change in illumination.

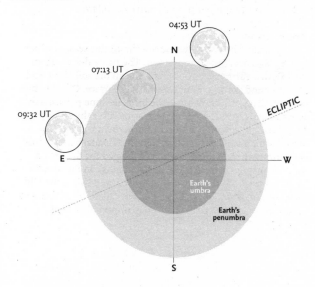

Calendar for March

03	08:54	Antares 0.4°S of the Moon
03	15:23	Last Quarter
03	18:01	Minor planet (3) Juno at opposition (mag. 8.7)
08	05:00	Mars (mag. 1.2) 3.5°N of the Moon
08	08:18 *	Neptune (mag. 8.0) 0.5°S of Mercury (mag. -1.5)
08	17:00	Venus (mag. -3.8) 3.3°N of the Moon
09	17:28	Saturn (mag. 1.0) 1.5°N of the Moon
10		Summer Time begins in US
10	07:04	Moon at perigee = 356,896 km (closest of the year)
10	09:00	New Moon
10	19:26	Neptune (mag. 8.0) 0.5°N of the Moon
11	02:31	Mercury (mag. -1.3) 1.0°N of the Moon
14–15		γ-Normid meteor shower maximum
14	01:02	Jupiter (mag. -2.1) 4.0°S of the Moon
14	11:35	Uranus (mag. 5.8) 3.4°S of the Moon
15	23:43	Aldebaran 9.9°N of the Moon
17	04:11	First Quarter
19	07:24	Pollux 1.4°N of the Moon
22	05:27	Regulus 3.6°S of the Moon
23	15:45	Moon at apogee = 406,294 km
24	22:34	Mercury at greatest elongation (18.7°E, mag. -0.2)
25	07:00	Full Moon
25	07:14	Penumbral lunar eclipse
26	20:23	Spica 1.4°S of the Moon
30	15:03	Antares 0.3°S of the Moon
31		Summer Time begins (BST in UK)

These objects are close together for an extended period around this time.

March 13–15 • *High in the west, the Moon passes Jupiter, the Pleiades and Aldebaran (as seen from central USA).*

March 13–16 • *The Moon passes Jupiter, the Pleiades and Aldebaran (this time as seen from London).*

March 27 • *The Moon and Spica are about thirty degrees above the western horizon (as seen from Sydney).*

March 31 • *The Moon is close to Antares. Sabik is lower and almost due east (as seen Sydney).*

M

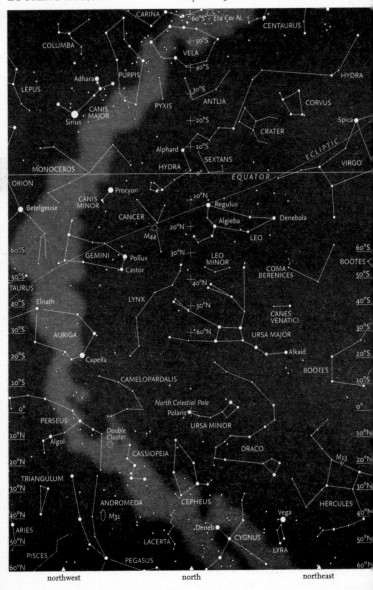

CARINA
60°S • Eta Car N.
CENTAURUS
COLUMBA
VELA
50°S
40°S
HYDRA
Adhara
PUPPIS
30°S
PYXIS
ANTLIA
CORVUS
LEPUS
CANIS
MAJOR
20°S
Sirius
CRATER
Spica
Alphard
10°S
SEXTANS
ECLIPTIC
MONOCEROS
HYDRA
0°
VIRGO
EQUATOR
ORION
10°N
Regulus
Betelgeuse
Procyon
Algieba
Denebola
CANIS
MINOR
CANCER
20°N
LEO
M44
Gemini
Pollux
30°N
LEO
MINOR
COMA
BERENICES
BOÖTES
60°S
50°S
Castor
TAURUS
50°S
40°N
40°S
40°S
Elnath
LYNX
50°N
CANES
VENATICI
30°S
30°S
AURIGA
60°N
URSA MAJOR
20°S
Capella
Alkaid
20°S
CAMELOPARDALIS
BOÖTES
10°S
10°S
North Celestial Pole
0°
0°
PERSEUS
Polaris
URSA MINOR
10°N
10°N
Algol
Double
Cluster
DRACO
M13
20°N
CASSIOPEIA
20°N
TRIANGULUM
HERCULES
30°N
30°N
ANDROMEDA
CEPHEUS
40°N
M31
Vega
40°N
ARIES
LACERTA
CYGNUS
Deneb
50°N
50°N
PISCES
PEGASUS
LYRA
60°N
60°N

northwest   north   northeast

March – Looking North

The highly distinctive (and widely recognized) constellation of **Ursa Major** with the distinctive asterism of the Plough (or Big Dipper) is now 'upside down' and near the zenith for observers in the far north, for whom it is particularly difficult to observe. At this time of year, it is high in the sky for anyone north of the equator. Only observers farther towards the south will find it lower down towards their northern horizon and reasonably easy to see. However, at 30°S, even the seven stars making up the main, easily recognized portion of the constellation are too low to be visible.

Auriga, with brilliant **Capella** (α Aurigae) is also very high on the opposite side of the meridian. The constellation of **Perseus** lies between it and **Andromeda** on the western side of the sky.

Ursa Minor, also with seven main stars, one of which is **Polaris**, the Pole Star, and the long constellation of **Draco** that winds around the Pole, are readily visible for anyone in the northern hemisphere, although, of course, Polaris is right on the horizon for anyone at the equator, and thus always lost to sight. **Cepheus** is near the meridian to the north, with **Cassiopeia**, to its west beginning to turn and resume its 'W' shape. The constellation of Andromeda is now diving down into the northwestern sky. In the east, beyond **Alkaid** (η Ursae Majoris), the final star in the 'tail' of Ursa Major, lies the top of **Boötes**. Farther to the south, most of **Hercules** and the 'Keystone' shape that forms the major portion of the body is visible.

On 13 March 1989 a major geomagnetic storm created a nine-hour disruption of Hydro-Quebec's electricity transmission system. The accompanying aurorae could be seen as far south as Texas and Florida. The geomagnetic storm was one of a number of incidents during a phase of major solar activity.

Observers at 50°N may occasionally be able to detect bright **Deneb** (α Cygni) and **Vega** (α Lyrae) skimming the horizon, together with portions of those particular constellations, although most of the time they will be lost in the extinction that occurs at such low altitudes.

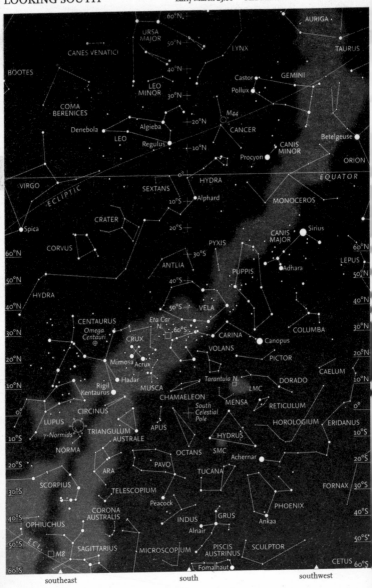

southeast   south   southwest

March – Looking South

The distinctive constellation of *Leo* is close to the meridian early in the night, clearly visible for anyone north of the equator. It is easily recognized, with bright **Regulus** (α Leonis) at the base of the 'backward question mark' of the asterism known as 'the Sickle', which lies north of Regulus. To the west of Leo is the inconspicuous constellation of *Cancer*, and still farther away from the meridian, the far more striking constellation of *Gemini*, with the bright stars *Castor* (α Geminorum), the closer to the Pole, and **Pollux** (β Geminorum). The constellation straddles the ecliptic, and Pollux may sometimes be occulted by the Moon (as may Regulus), although no such occultation occurs in 2024. Castor is remarkable in that even a fairly small telescope will show it as consisting of three stars (two fairly bright, and one fainter). However, more detailed investigation reveals that each of those stars is actually a double, so the whole system consists of no fewer than six stars.

Below Cancer is the very distinctive asterism of the 'Head of Hydra', consisting of five (or six) stars, that is the western end of the long constellation of **Hydra**, the largest constellation in the sky, that runs far towards the east, roughly parallel to the ecliptic. **Alphard** (α Hydrae) is south, and slightly to the west of Regulus in Leo and is relatively easy to recognize as it is the only fairly bright star in that region of the sky. North of Hydra and between it and the ecliptic and the constellation of *Virgo* are the two constellations of *Crater* and *Corvus*. Farther west, the small constellation of *Sextans* lies between Hydra and Leo.

Farther south, the Milky Way runs diagonally across the sky, and the constellation of *Vela* straddles the meridian. Slightly farther south is the constellation of *Carina*, with, to the west, brilliant *Canopus* (α Carinae), which lies below the constellation of *Puppis*, which is itself between Vela and *Canis Major* in the west.

Crux (the Southern Cross) is southeast of Carina and the two principal stars of *Centaurus*, *Rigil Kentaurus* (α Centauri) and *Hadar* (β Centauri). The *Large Magellanic Cloud* (LMC) lies west of these stars, on the other side of the meridian.

Caroline Herschel

Caroline Herschel, the famous sister of William Herschel, his helpmate and a significant astronomer in her own right, was born in Hannover, in Germany on 16 March 1750. Caroline discovered eight comets, having piority over five of these. The total included periodic comet 35P/Herschel–Rigollet and the rediscovery of Comet 2P/Encke. Because of her work, she became the first woman to earn a salary as a scientist. She was the first woman to publish in the Royal Society's Philosophical Transactions and, together with Mary Sommerville, was the first to be nominated as an Honorary Fellow of the Royal Astronomical Society. On William's death in 1822, she returned to Hannover and died there on 9 January 1848, aged 97. The crater Caroline Herschel, named in her honour, is a small, bowl-shaped crater (7 km in diameter) in the northwestern part of the lunar Mare Imbrium.

Next page: *The crater Caroline Herschel, named after the astronomer, is just 7 kilometres in diameter and is located in the Mare Imbrium on the Moon.*

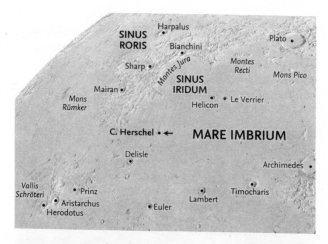

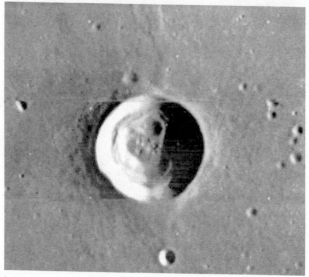

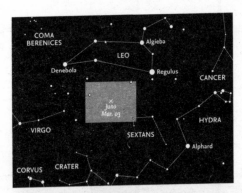

Minor planet **(3) Juno** comes to opposition on March 3. The lighter grey box is shown in more detail below. The white cross marks the position of Juno at the day of its opposition.

*The path of the minor planet **(3) Juno** around its opposition on March 3 (mag. 8.7). Background stars are shown down to magnitude 9.5.*

April

April – Introduction

The principal astronomical event in April 2024 is the total solar eclipse of April 8, with maximum eclipse over Mexico. The path of totality enters the United States, and runs across many of the states on the eastern seaboard, before entering Canada in Nova Scotia and ending over the North Atlantic Ocean. The full path is shown in the map on page 103.

Three meteor showers are active during the month. Two are best seen from the northern hemisphere, with one significant southern shower. The one moderate, northern shower is the

The charts show the location of the April Lyrids radiant (top), the π-Puppids radiant (middle) and the η-Aquariids radiant (bottom).

Lyrids (often called the **April Lyrids** to distinguish them from several minor showers that originate in the constellation at various times during the year). In 2024 the shower begins on April 14, one day before First Quarter, and comes to a weak maximum of 18 meteors per hour on April 22–23, just before and at Full Moon, so conditions are generally unfavourable.

There is another, stronger, northern shower, with a possible hourly rate of 50 meteors per hour. This is the *η-Aquariid* shower. This begins on April 19, four days before Full Moon, and peaks on May 6, two days before New Moon, continuing well into May (May 28). So conditions for observing this shower are poor at the beginning, improve around maximum, but then deteriorate again towards Full Moon. The η-Aquariid shower, like the Orionids in October, is the result of particles left in orbit behind the famous comet 1P/Halley. This is a moderately predictable shower, and the maximum rate is between 40 and 50 meteors per hour. The radiant for this shower is close to the 'Y'-shaped asterism known as the 'Water Jar' in Aquarius.

The southern meteor shower, the *π-Puppids*, begins in April. This shower starts to be active on April 15, at

On 5 April 2019, the Japanese spacecraft *Hayabusa2* fired an impactor at the asteroid *Ryugu*, creating a crater from which samples were later obtained.

First Quarter and lasts until April 28. The maximum is on the night of April 23–24, at and immediately after Full Moon, so the overall performance and the maximum in particular is highly unfavourable.

The planets
Mercury is too close to the Sun to be visible this month. *Venus* is similarly placed. *Mars* is moving towards the Sun, but may be glimpsed in the morning sky early in the month. *Jupiter* (mag. -2.0) is in *Aries*, crossing into *Taurus* at the end of the month. *Saturn* (mag. 1.1 to 1.2) is moving slowly in *Aquarius*. *Uranus* (mag. 5.8) remains in *Aries* and *Neptune* (mag. 8.0 to 7.9) is in *Pisces*.

Sunrise and sunset

City	Date	Sunrise	Sunset
Buenos Aires, Argentina			
	Apr. 01	10:06	21:48
	Apr. 30	10:29	21:12
Cape Town, South Africa			
	Apr. 01	04:58	16:41
	Apr. 30	05:20	16:06
London, UK			
	Apr. 01	05:36	18:34
	Apr. 30	04:35	19:23
Los Angeles, USA			
	Apr. 01	13:40	02:13
	Apr. 30	13:05	02:36
Nairobi, Kenya			
	Apr. 01	03:34	15:39
	Apr. 30	03:28	15:32
Sydney, Australia			
	Apr. 01	20:08	07:51
	Apr. 30	20:30	07:16
Tokyo, Japan			
	Apr. 01	20:27	09:02
	Apr. 30	19:50	09:26
Washington, DC, USA			
	Apr. 01	10:53	23:32
	Apr. 30	10:11	24:00
Wellington, New Zealand			
	Apr. 01	18:37	06:13
	Apr. 30	19:08	05:29

NB: the times given are in Universal Time (UT)

The Moon's phases and ages

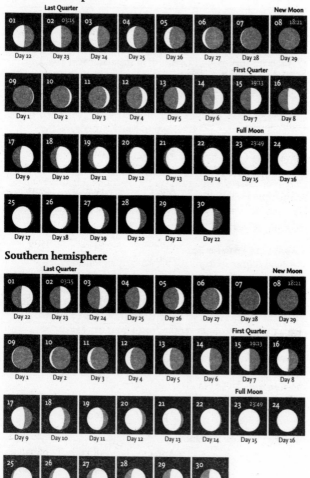

Northern hemisphere

Last Quarter

01	02 03:15	03	04	05	06	07	08 18:21
Day 22	Day 23	Day 24	Day 25	Day 26	Day 27	Day 28	Day 29

New Moon

First Quarter

09	10	11	12	13	14	15 19:13	16
Day 1	Day 2	Day 3	Day 4	Day 5	Day 6	Day 7	Day 8

Full Moon

17	18	19	20	21	22	23 23:49	24
Day 9	Day 10	Day 11	Day 12	Day 13	Day 14	Day 15	Day 16

25	26	27	28	29	30
Day 17	Day 18	Day 19	Day 20	Day 21	Day 22

Southern hemisphere

Last Quarter

01	02 03:15	03	04	05	06	07	08 18:21
Day 22	Day 23	Day 24	Day 25	Day 26	Day 27	Day 28	Day 29

New Moon

First Quarter

09	10	11	12	13	14	15 19:13	16
Day 1	Day 2	Day 3	Day 4	Day 5	Day 6	Day 7	Day 8

Full Moon

17	18	19	20	21	22	23 23:49	24
Day 9	Day 10	Day 11	Day 12	Day 13	Day 14	Day 15	Day 16

25	26	27	28	29	30
Day 17	Day 18	Day 19	Day 20	Day 21	Day 22

A

The Moon

The Moon in April
On April 8, at New Moon, there is a total solar eclipse, with the track covering Mexico, the eastern states of the USA and eastern Canada (see the facing pages). The Moon then passes the *Pleiades* and by April 12 is 10° north of *Aldebaran* in *Taurus*. It is 1.5° south of *Pollux* by April 15, and 3.5° north of *Regulus*, between it and *Algieba* on April 18. Before Full Moon on April 23, it is 1.4° north of *Spica* in *Virgo* and 0.3° north of *Antares* by April 26.

Pink Moon
The Full Moon in April is known in North America as the Pink Moon, from the pink flowers – phlox – that bloom in the early spring. Other names for this Full Moon used by Native American tribes include Sprouting Grass Moon, Fish Moon and Hare Moon. On the other side of the Atlantic there is the Old English/Anglo-Saxon name of Egg Moon. In Europe generally it is sometimes known as the Paschal Moon because it is used to calculate the date for Easter.

8 April 2024 – solar eclipse
The total eclipse of April 8 is particularly significant for obvservers (see the maps on the facing page). The path of totality passes across Mexico, entering the United States in Texas, then running across the southern and eastern states. It leaves the eastern seabord in Canadian Nova Scotia, crossing Newfoundland, to end over the North Atlantic. Maximum eclipse actually occurs in Mexico at 18:17, where the duration is 4 minutes 28.1 seconds.

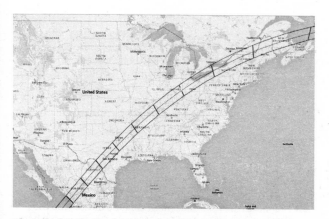

The path of totality for the eclipse of April 8 as it passes across Mexico, and the eastern states of the United States.

On 8 April 2024 totality begins in the middle of the Pacific Ocean. The path then crosses Mexico and enters the United States. It ends over the North Atlantic.

Calendar for April

02	03:15	Last Quarter
06	03:51	Mars (mag. 1.2) 2.0°N of the Moon
06	09:24	Saturn (mag. 1.1) 1.2°N of the Moon
07	08:11	Neptune (mag. 8.0) 0.4°N of the Moon
07	16:38	Venus (mag. -3.8) 0.4°S of the Moon
07	17:51	Moon at perigee = 358,849 km
08	18:17	Total solar eclipse (Mexico/USA)
08	18:21	New Moon
09	01:24	Mercury (mag. 4.9) 2.2°N of the Moon
10	21:09	Jupiter (mag. -2.0) 4.0°S of the Moon
10	23:51	Uranus (mag. 5.8) 3.6°S of the Moon
12	08:48	Aldebaran 10°S of the Moon
14–30		April Lyrid meteor shower
15–28		π-Puppid meteor shower
15	14:25	Pollux 1.5°N of the Moon
15	19:13	First Quarter
18	11:56	Regulus 3.5°S of the Moon
18	23:00 *	Venus (mag. -3.8) 2.0°S of Mercury (mag. 3.9)
19–May 28		η-Aquariid meteor shower
20	02:10	Moon at apogee = 405,623 km
22–23		April Lyrid meteor shower maximum
23–24		π-Puppid meteor shower maximum
23	02:44	Spica 1.4°S of the Moon
23	23:49	Full Moon
26	20:39	Antares 0.3°S of the Moon

* These objects are close together for an extended period around this time.

April 10–11 • *After sunset, the narrow crescent Moon passes Jupiter and the Pleiades (as seen from London).*

April 18–19 • *The waxing gibbous Moon passes between Regulus and Algieba (as seen from London).*

April 22 • *High in the southeast, the Moon and Spica are side by side (as seen from central USA).*

April 27 • *After sunset, Jupiter is eight degrees below the Pleiades. Aldebaran is farther west (as seen from central USA).*

April – Looking North

Ursa Major is still high overhead for most northern-hemisphere observers, with **Boötes** and bright **Arcturus** (α Boötis), the star indicated by the 'arc' of the 'tail' of Ursa Major, to its northeast. The small circlet of **Corona Borealis**, with its single bright star of **Alphecca** (α CrB) is still farther to the east. **Cassiopeia** has now swung round, and is almost on the meridian to the north, below **Polaris** and northwest of **Ursa Minor**. **Cepheus** is beginning to climb higher and the head of **Draco** is almost due east of Polaris, but slightly farther south than the two 'Guards' in Ursa Minor, **Kochab** (β UMi) and **Pherkad** (γ UMi). On the other side of the meridian to the northwest lies the constellation of **Auriga**, with bright **Capella**. The inconspicuous constellation of **Camelopardalis** lies between Polaris, Auriga and **Perseus**. The stars of the Milky Way run through Auriga, Perseus, and Cassiopeia towards the much more densely populated regions in **Cygnus** and farther south. Nearly the whole of Cygnus is visible above the horizon in the northeast, where the small constellation of Lyra, with the bright star **Vega**, is clearly seen with **Hercules** above it.

The constellation of Perseus is beginning to descend into the northwest, following the constellation of **Andromeda**, which, for most observers, has now disappeared below the northwestern horizon.

A

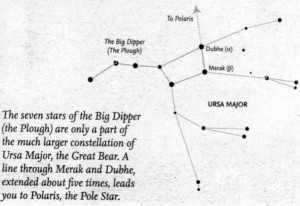

To Polaris

The Big Dipper
(The Plough)

Dubhe (α)

Merak (β)

URSA MAJOR

The seven stars of the Big Dipper (the Plough) are only a part of the much larger constellation of Ursa Major, the Great Bear. A line through Merak and Dubhe, extended about five times, leads you to Polaris, the Pole Star.

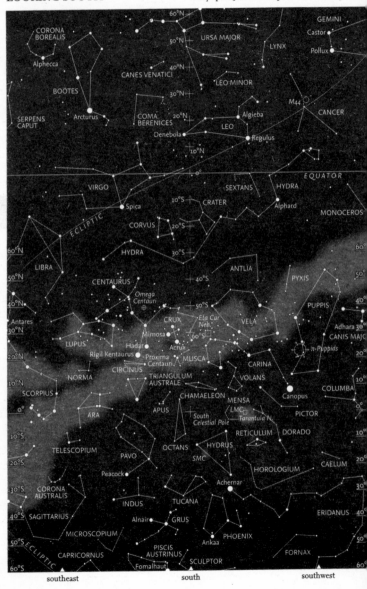

southeast   south   southwest

April – Looking South

The constellation of **Leo** is still conspicious in the south, although now it is **Denebola** (β Leonis) rather then **Regulus** that is close to the meridian. **Boötes** and **Arcturus** (α Boötis), the brightest star in the northern hemisphere, are high in the southeast. Beyond Leo, along the ecliptic to the east, the zodiacal constellation of **Virgo** is clearly visible, although **Spica** (α Virginis) is the only really bright star in the constellation.

Farther south, **Crux** is now nearly 'upright' and close to the meridian, with **Rigil Kentaurus** and **Hadar** (α and β Centauri, respectively), conspicuous to its southeast. Rigil Kentaurus is a multiple system, with two stars fairly easily visible with a small telescope. A third star in the system, known as **Proxima Centauri**, is much fainter, and is actually the closest star to the Solar System at a distance of about 4.3 light-years. In recent years, it has been found to host the closest known exoplanet, which orbits Proxima every 11.2 days.

The stars of **Vela** and **Carina** lie to the west of Crux and the meridian, while much farther south, and out to the west is **Canopus** (α Carinae) the second-brightest star in the whole sky after Sirius in Canis Major. Even farther south, and right on the horizon for people at 30°S (roughly the latitude of Sydney in Australia) is **Achernar** (α Eridani), the main star in the long, winding constellation of **Eridanus** (the River), which begins near the star Rigel in Orion, far to the north.

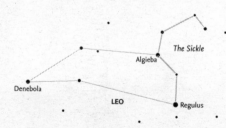

The distinctive constellation of Leo, with Regulus and 'The Sickle' on the west. Algieba (γ Leonis), north of Regulus, appearing double, is a multiple system of four stars.

The Moon Illusion

Frequently, when the Moon is seen rising or setting, and it is close to the horizon, it appears absolutely enormous. It seems far larger than when it is high in the sky. In fact, this is an optical illusion – it is known as the 'Moon Illusion'. Our eyes are playing tricks. In reality, the Moon is exactly the same size wherever it is in the sky. Try covering it with your finger, held at arm's length. Any finger is more than large enough to cover the Moon. (In fact, the Moon is about 30 minutes across, and a finger is about twice that, about 1 whole degree.) There have been many attempts over the years to explain the Moon Illusion, but it seems that the brain automatically compares the size of the Moon with the distant horizon, and assumes that it is at the same distance – whereas it is, of course, much farther away. The first person to correctly explain the effect was the Arab expert on optics, Ḥasan Ibn al-Haytham, known as Alhazen, in a book written in 1011–1021.

When the Moon is seen against distant objects, the brain is tricked into thinking it is much larger than its true size.

May

May – Introduction

The *η-Aquariid* meteor shower comes to maximum on May 6. This shower, like the Orionids in October, is associated with debris from Comet 1P/Halley. Maximum in 2024 is on Day 28 of the lunation, just before New Moon, so conditions are very favourable. The radiant is close to the celestial equator, so conditions for northern observers are poor, even though the shower does occasionally produce bright meteors that leave persistent trains.

In May 2020, it was announced that two new southern meteor streams had been discovered. These streams are apparently associated with long-period comets. The first of these new meteor streams is the *γ-Piscid Austrinids*, which appear to peak early in the morning of May 15 and may well be an annual shower. This date is on Day 7 of the lunation, at First Quarter in 2024, so observing conditions are not particularly good. The second southern shower, is the *σ-Phoenicids* and this appears to peak slightly later on the same date. The data gathered over the preceding years suggest that the rate for this shower may vary considerably from year to year. Both new showers are known from extremely low numbers of meteors only (15 and 14, respectively) and have yet to be recognized as 'official' meteor showers.

On 25 May 2008, NASA's *Phoenix* spacecraft landed in the northern polar region of Mars.

The planets

Mercury (mag. 0.4) reaches greatest elongation west in the morning sky on May 9. By the end of the month on May 31 (at mag. -0.7) it is 1.4° south of **Uranus** (mag. 5.8). **Venus** is nearing conjunction with the Sun (June 4), and so is invisible this month. **Mars** is moving towards the Sun, so is also invisible. **Jupiter** (mag. -2.0), now in **Taurus**, is moving towards the Hyades. **Saturn** is moving slowly east in **Aquarius**. **Uranus** (mag. 5.8) remains in **Aries** and **Neptune** (mag. 7.9) in **Pisces**.

M

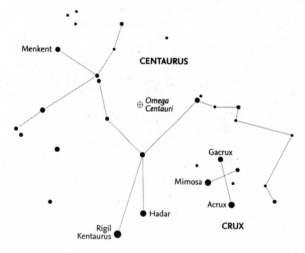

In May, the constellation of Centaurus is well placed for observing from the tropical regions and farther south. It contains the largest and finest globular star cluster in the sky, Omega Centauri. Crux, the Southern Cross is between the front and the hind legs of the Centaur.

Sunrise and sunset

City	Date	Sunrise	Sunset
Buenos Aires, Argentina			
	May 01	10:29	21:11
	May 31	10:51	20:51
Cape Town, South Africa			
	May 01	05:21	16:05
	May 31	05:42	15:45
London, UK			
	May 01	04:33	19:24
	May 31	03:50	20:08
Los Angeles, USA			
	May 01	13:04	02:36
	May 31	12:43	02:58
Nairobi, Kenya			
	May 01	03:28	15:32
	May 31	03:29	15:32
Sydney, Australia			
	May 01	20:30	07:15
	May 31	20:51	06:55
Tokyo, Japan			
	May 01	19:49	09:27
	May 31	19:27	09:50
Washington, DC, USA			
	May 01	10:11	24:00
	May 31	09:45	00:26
Wellington, New Zealand			
	May 01	19:08	05:28
	May 31	19:37	05:01

NB: the times given are in Universal Time (UT)

The Moon's phases and ages

Northern hemisphere

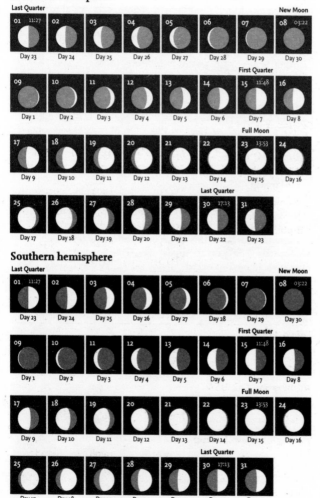

Last Quarter

							New Moon
01 11:27	02	03	04	05	06	07	08 03:22
Day 23	Day 24	Day 25	Day 26	Day 27	Day 28	Day 29	Day 30

First Quarter

09	10	11	12	13	14	15 11:48	16
Day 1	Day 2	Day 3	Day 4	Day 5	Day 6	Day 7	Day 8

Full Moon

17	18	19	20	21	22	23 13:53	24
Day 9	Day 10	Day 11	Day 12	Day 13	Day 14	Day 15	Day 16

Last Quarter

25	26	27	28	29	30 17:13	31
Day 17	Day 18	Day 19	Day 20	Day 21	Day 22	Day 23

Southern hemisphere

Last Quarter

							New Moon
01 11:27	02	03	04	05	06	07	08 03:22
Day 23	Day 24	Day 25	Day 26	Day 27	Day 28	Day 29	Day 30

First Quarter

09	10	11	12	13	14	15 11:48	16
Day 1	Day 2	Day 3	Day 4	Day 5	Day 6	Day 7	Day 8

Full Moon

17	18	19	20	21	22	23 13:53	24
Day 9	Day 10	Day 11	Day 12	Day 13	Day 14	Day 15	Day 16

Last Quarter

25	26	27	28	29	30 17:13	31
Day 17	Day 18	Day 19	Day 20	Day 21	Day 22	Day 23

M

The Moon

The Moon in May

On May 3, the Moon is 0.8° south of **Saturn** (mag. 1.2) and the next day is 0.3° south of much fainter **Neptune** (mag. 7.9). By May 5 it is 0.2° north of **Mars** (mag. 1.0). On May 6 it is 3.8° north of **Mercury** (mag. 0.6) and the next day 3.5° north of much brighter **Venus** (mag. -3.9). At New Moon on May 8, it is 3.6° north of faint **Uranus** (mag. 5.8) and a few hour later 4.3° north of **Jupiter** (mag. -2.0). On May 9 it is 9.9° north of **Aldebaran** in **Taurus**. On May 12 the Moon is 1.6° south of **Pollux**. At First Quarter on May 15, it is 3.5° north of **Regulus**. By May 20, the Moon is 1.4° north of **Spica** in **Virgo**. On May 24, the Moon (just past Full) is 0.4° north of **Antares**. By May 31, the Moon is 0.4° south of **Saturn**.

C.A. Hartwig

The famous German astronomer C.A. Hartwig died on 3 May 1923. Hartwig, who was born on 14 January 1851 in Frankfurt, was noted for his study of variable stars. He discovered the famous supernova (SN 1885A or S Andromedae) in the Andromeda Galaxy on 20 August 1885. This was the first supernova to be discovered outside the Milky Way Galaxy. ('Tycho's Supernova', B Cassiopeiae, of 1572 and 'Kepler's

Carl Ernst Albrecht Hartwig.

Supernova', SN 1604, of 1604 were both in our own galaxy.) In 1887 Hartwig became director of the Remeis Observatory at Bamberg.

A lunar crater is named after him. It lies on the western limb of the Moon and is about 80 kilometres across and lies on the outer ejecta blanket surrounding the Mare Orientale basin. The ejecta from that impact have greatly modified the crater and overlie most of the eastern rim and the floor. Only part of the western rim appears intact. There is a smaller crater on the crater floor towards this part of the rim.

A crater in the southern hemisphere of Mars is also named after the astronomer. This crater is slightly larger than the lunar crater, being about 105 kilometres in diameter.

M

The lunar crater Hartwig is close to the edge of the Moon, to the west of the Grimaldi basin.

Calendar for May

01	11:27	Last Quarter
03	22:32	Saturn (mag. 1.2) 0.8°N of the Moon
04	18:55	Neptune (mag. 7.9) 0.3°N of the Moon
05	02:25	Mars (mag. 1.0) 0.2°S of the Moon
05	22:04	Moon at perigee = 363,163 km
06		η-Aquariid meteor shower maximum
06	08:25	Mercury (mag. 0.6) 3.8°S of the Moon
07	16:03	Venus (mag. -3.9) 3.5°S of the Moon
08	03:22	New Moon
08	12:51	Uranus (mag. 5.8) 3.6°S of the Moon
08	18:14	Jupiter (mag. -2.0) 4.3°S of the Moon
09	18:54	Aldebaran 9.9°S of the Moon
09	21:29	Mercury at greatest elongation (26.4°W, mag. 0.4)
12	22:54	Pollux 1.6°N of the Moon
15	11:48	First Quarter
15	19:24	Regulus 3.5°S of the Moon
17	18:59	Moon at apogee = 404,640 km
19	15:06	Minor planet (2) Pallas at opposition (mag. 9.0)
20	10:03	Spica 1.4°S of the Moon
23	13:53	Full Moon
24	03:10	Antares 0.4°S of the Moon
30	17:13	Last Quarter
31	01:00 *	Uranus (mag. 5.8) 1.4°N of Mercury (mag. -0.7)
31	08:09	Saturn (mag. 1.2) 0.4°N of the Moon

These objects are close together for an extended period around this time.

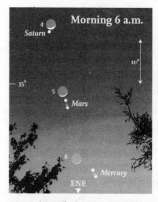

May 4–6 • *The crescent Moon passes Saturn, Mars and Mercury (as seen from Sydney).*

May 12–13 • *In the northwest, the Moon passes the Twin Stars, Pollux and Castor (as seen from Sydney).*

May 15 • *Just past First Quarter, the Moon is between Regulus and Algieba (as seen from London).*

May 24 • *In the early morning, the Full Moon is close to Antares, low in the south-southwest (as seen from London).*

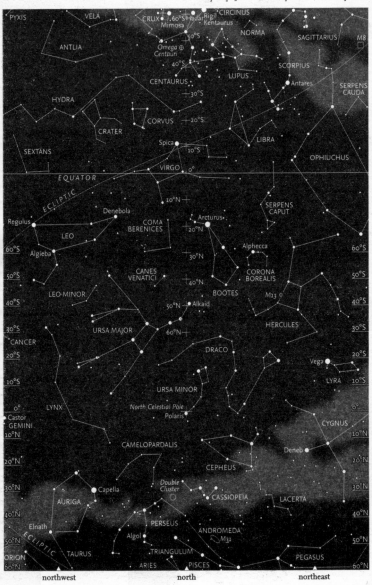

northwest   north   northeast

May – Looking North

The constellation of **Cassiopeia** is now low over the northern horizon, to the east of the meridian, to observers at mid-northern latitudes. To its west, the southern portions of both **Perseus** and **Auriga** are becoming difficult to observe as they become closer to the horizon, although the **Double Cluster**, between Perseus and Cassiopeia, is still clearly visible.

High above, **Ursa Major** has started to swing round to the west and the stars that form the extended portion towards the south (the 'legs' and 'paws' of the 'bear') are becoming easier to see. **Alkaid** (η Ursae Majoris), at the end of the 'tail', is almost exactly at the zenith for observers at 50° North. The whole of the constellations of **Cepheus**, **Draco**, **Lyra** and **Ursa Minor** are easy to see, together with the inconspicuous constellation of **Camelopardalis**. For those people with the keenest eyesight, they can even make out the long, straggling line of faint stars forming the constellation of **Lynx** in the northwestern sky, beyond the extended portion of Ursa Major. These faint constellations become difficult to detect during the brighter nights towards the end of the month.

Also high in the sky for northern observers are the constellations of **Boötes**, with bright **Arcturus**, and **Hercules**, with its clearly visible 'Keystone' of four stars, on one side of which is **M13**, the finest globular cluster in the northern sky. The small, but striking arc of stars forming the constellation of **Corona Borealis** lies between Hercules and Boötes. It has just a single bright star, **Alphecca** (α Coronae Borealis).

Much of **Cygnus** is clearly seen in the east, as are two (**Deneb** and **Vega**) of the three stars that form the Summer Triangle. The third star, **Altair** in Aquila, begins to climb above the horizon late in the night and later in the month. Across the Milky Way, between Cassiopeia and Cygnus, and below Cepheus, is the tiny zig-zag of stars that forms the small, and often ignored, constellation of **Lacerta**.

M

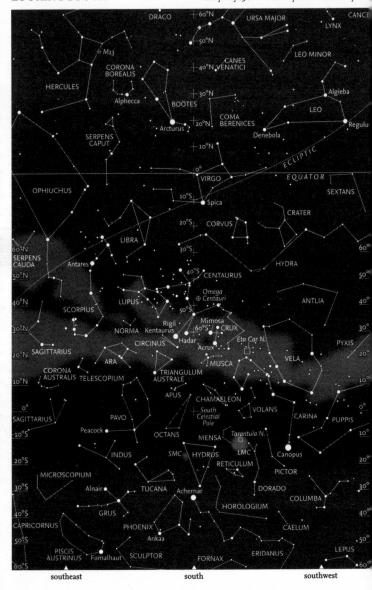

southeast　　　　　　　　　　south　　　　　　　　　　southwest

May – Looking South

Spica (α Virginis) in the sprawling constellation of **Virgo** is now close to the meridian. The constellation of **Boötes**, with brilliant *Arcturus* (the brightest star in the northern hemisphere of the sky) is higher above and slightly to the east, clearly visible to observers in the north. Following Virgo across the sky is the rather undistinguished constellation of **Libra** and, following it, and rising in the east, **Scorpius**, with the red supergiant star *Antares* and the distinctive line of stars, running south and ending in the 'sting'. (As Orion sinks in the west, so Scorpius rises in the east. This recalls one of the many legends about Orion: with his being pursued by a scorpion, sent to distract him while fighting.) Above Libra and Scorpius, the large constellation of **Ophiuchus** is beginning to climb higher in the sky. It lies between the two portions of **Serpens** (the only constellation to be divided into two parts). **Serpens Caput** (the Head of the Serpent) lies west of Ophiuchus, between it and Boötes, and **Serpens Cauda** (the Tail of the Serpent) is to the east, between Ophiuchus and the Milky Way.

Farther south, between Scorpius and the two brightest stars in **Centaurus** (*Rigil Kentaurus* and *Hadar*), lies the constellation of **Lupus**, just east of the meridian and lying along the Milky Way. **Crux** is now more-or-less 'upright', with the small constellation of **Musca** below it. Below the bright pair of stars in Centaurus (α and β Centauri) is the constellation of **Triangulum Australe**, a much larger and more striking constellation than its counterpart (Triangulum) in the north. Lying between Rigil Kentaurus and Triangulum Australe is the very tiny and indistinct constellation of **Circinus**.

The stars of **Vela** and **Carina** are becoming lower in the southwest, following the constellation of **Puppis** and brilliant *Canopus* (α Carinae) down towards the horizon. Puppis and the **Large Magellanic Cloud** (LMC) are close to the horizon for anyone at the latitude of 30°S (about the latitude of Sydney in Australia), where *Achernar* (α Eridani) is now too low to be seen.

There are several, small, relatively faint constellations in this part of the sky. The most distinct is probably **Pavo**, with its single bright star (α Pavonis), known as **Peacock**. Other constellations are **Apus**, **Chamaeleon**, **Octans** (which actually includes the South Celestial Pole), **Mensa** and **Volans**.

M

Comet Tempel 1

On 26 May 2024 Comet Tempel 1 passes so close to Jupiter that it may break up under the planet's gravitational forces. Comet Tempel 1 (officially 9P/Tempel) was discovered in 1867. It has an orbital period of just 5.56 years, and occasionally passes close to Jupiter and as a result, suffers a change in orbit, orbital period and perihelion. For this reason, it was unsuccessfully searched for (and not found) in 1898 and 1905.

The comet was the target for the *Deep Impact* space mission, with impact occurring on 4 July 2005. The result of the impact could not be photographed by the *Deep Impact* spacecraft itself, because of the cloud of debris raised by the impact, although observations were made from Earth and space. The spectrometer on the probe did, however, detect silicates, carbonates, metal sulphides and polycyclic aromatis hydrocarbons (PAH). Water was also found from a layer underlying the crust.

A second mission to the comet was authorized (the New Exploration of Tempel 1 – NExT), using the existing *Stardust* spacecraft, which had studied Comet Wild 2 (81P/Wild) and collected a small number of sample particles in 2004. The spacecraft was placed in an orbit that took it to 181 km of the comet. The crater formed by the impactor was estimated to be 150 m across and about 28 m deep.

Facing page, top:
Comet Tempel 1, as imaged by the Deep Impact *spacecraft.*

Facing page, bottom:
Comet Tempel 1, as imaged by the the Stardust *spaceprobe.*

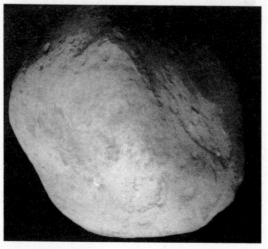

June

June – Introduction

June is a very quiet month for astronomical activity. There are no major meteor showers and because of the light nights, normal astronomy is difficult for northern observers. Time to turn one's attention to noctilucent clouds (for northern observers) and to 'daytime astronomy' – the study of the Sun.

There is a commonly-held view that the weather in June is hot because the Earth is then closest to the Sun, but this is completely wrong. The Earth's orbit is not a true circle, but an ellipse. In 2024 the distance varies from 147,100,632.54 km (0.983306994 AU) on January 3 (at perihelion) to 152,099,968.24 km (1.016725489 AU) on July 5 (at aphelion) in 2024. So the Earth is actually closest to the Sun in January. The winter solstice (for northern-hemisphere residents) when the Sun is lowest in the sky, is on December 21 in 2024 (December 22 in 2023). At the summer solstice (winter for those in the southern hemisphere) on June 21, it is actually farther from the Sun, reaching its farthest point on July 6.

The cause of the seasons, with warm summers and cold winters, is solely caused by the tilt of the Earth's axis to the plane of its orbit around the Sun – the ecliptic. As such, the northern hemisphere is most strongly tilted towards the Sun in June and July, and the southern hemisphere in November and December. At the equinoxes (March 20 and September 22 in 2024), the Earth's axis is tilted so that the planet is 'side on' to the Sun, so day and night are of (approximately) equal length and the Sun's heat is evenly distributed between the two hemispheres.

On 30 June 1908 an extremely violent airburst occurred in the Podkamennaya Tunguska (Stony Tunguska) River region of Siberia. The explosion flattened an estimated 80 million trees over an area of about 2,150 km^2. There is controversy over whether the impacting body was asteroidal or cometary in nature. It was most probably an airburst at an altitude of between 5 and 10 kilometres, caused by a stony meteoroid some 50–60 m in diameter.

The planets

Mercury moves from the morning to the evening sky and is mag. -1.2 and 4.7° north of the Moon on June 5 (one day before New Moon). **Venus** is very close to the Sun in **Gemini**. **Mars** (mag. 1.0) moves from **Pisces** into **Aries**. **Jupiter** (mag. -2.0) is in **Taurus**, 4.7° south of the Moon on June 5 (one day before New Moon). **Saturn** (mag. 1.1) is moving slowly east in **Aquarius**. **Uranus** (mag. 5.8) is now in **Taurus** and **Neptune** (mag. 7.9) remains in **Pisces**.

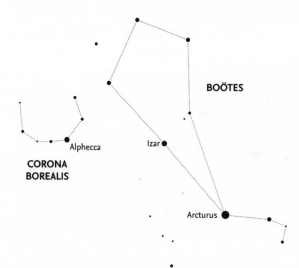

In June, the constellations of Boötes (with Arcturus) and Corona Borealis (with Alphecca) are well placed for observation. Arcturus has an orange tint and is the brightest star in the northern celestial hemisphere, at magnitude -0.05.

Sunrise and sunset

City	Date	Sunrise	Sunset
Buenos Aires, Argentina			
	Jun. 01	10:52	20:51
	Jun. 30	11:01	20:53
Cape Town, South Africa			
	Jun. 01	05:43	15:45
	Jun. 30	05:52	15:47
London, UK			
	Jun. 01	03:49	20:09
	Jun. 30	03:47	20:22
Los Angeles, USA			
	Jun. 01	12:43	02:59
	Jun. 30	12:45	03:09
Nairobi, Kenya			
	Jun. 01	03:29	15:32
	Jun. 30	03:35	15:38
Sydney, Australia			
	Jun. 01	20:52	06:54
	Jun. 30	21:01	06:57
Tokyo, Japan			
	Jun. 01	19:26	09:51
	Jun. 30	19:29	10:01
Washington, DC, USA			
	Jun. 01	09:45	00:27
	Jun. 30	09:46	00:38
Wellington, New Zealand			
	Jun. 01	19:38	05:01
	Jun. 30	19:48	05:01

NB: the times given are in Universal Time (UT)

The Moon's phases and ages

Northern hemisphere

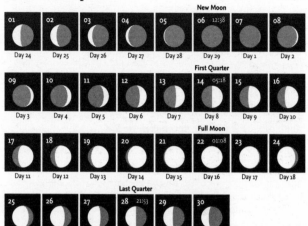

Southern hemisphere

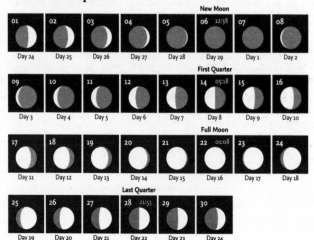

J

The Moon

The Moon in June

On June 1, the Moon passes extremely close and to the south of
Neptune, but there is no occultation. The next day it is 2.4° north
of **Mars**. On June 5, the day before New Moon, it passes 3.7° north
of **Uranus** (mag. 5.8) and later that day 4.7° north of **Jupiter**
(mag. -2.0). Later again, that same day, it is 4.7° north of **Mercury**
(mag. -1.2). It is 9.9° north of **Aldebaran** on June 6 (at New Moon).
By June 9 it is 1.7° south of **Pollux** and by June 12, 3.3° north of
Regulus in **Leo**. On June 16 it is 1.2° north of **Spica**. By June 20,
the Moon is 0.3° south of **Antares**. By June 27, the Moon is 0.1°
south of **Saturn** and, the next day (June 28) it is 0.3° north of
Neptune (mag. 7.9).

Strawberry Moon

June has always been noted for strawberries, and this appears in
the name of the Full Moon this month (on June 22). This name
was used on both sides of the Atlantic. The Choctaw tribe of
southeastern America had different names for Full Moons that
occurred in early or late June. In early June it was 'Moon of the
peach' and in late June it was 'Moon of the crane'. Other names are
'Hot Moon' and 'Rose Moon' and in the Old World, 'Mead Moon'.

Transits of Venus

Transits of Venus across the face of the Sun are actually quite rare.
They occur in pairs with an interval of about 8 years. The last was
in 2012 (preceded by the one in 2004) and the next will be in 2117.
Transits are rare because of the requirements that the orbits of
the Earth and Venus should be in line, and also that the orbital
periods should combine. Obviously, the timing and the location
of the two planets and their orbits need to be in the correct
alignment for a transit to occur. This recurs in a pattern of some
243 years, with long intervals of over 121 and 105 years between
pairs of transits.

The Persian astronomer Avicenna (980 to 1037 CE) claimed to
have seen Venus on the face of the Sun. There was a transit in 1032
CE, but his observation is disputed and he may have mistaken a
sunspot for the planet.

The first to predict a transit was Johannes Kepler, who, in 1627, predicted the transit of 1631. But this was not observed. The first confirmed observations were of the transit of 4 December 1639 by Jeremiah Horrocks and William Crabtree.

In 1691, the famous astronomer Edmond Halley predicted a transit of Venus in 1761 and suggested that this could be used to refine the scale of the Solar System, and primarily determine the value of the Astronomical Unit (the distance between the Earth and the Sun) by making observations from widely separated points on the surface of the Earth. Although Halley died in 1742, his predictions were respected and numerous expeditions were mounted, some of which were successful. The second transit of the pair, in 1769, was also observed from many different locations, although timings were complicated by what has come to be known as the 'black drop effect', a dark bar that appeared at the edge of the planet's disc at the various contacts. This was long thought to be caused by the thick atmosphere of Venus, but has since been shown to be an optical illusion crested by turbulence in the Earth's own atmosphere. In 1771, the French astronomer Jérôme Lalande used the observations of 1761 and 1769 to calculate the Astronomical Unit's value as 153 million km. (This may be compared with the modern value of 149,597,870 km.)

Transits also occurred in 1874 and 1882, and the latest pair in 2004 and 2012.

The transit of Venus on 8 June 2004 as photographed from Degrania A, in Israel.

Calendar for June

01	02:54	Neptune (mag. 7.9) 0.0°N of the Moon
02	07:16	Moon at perigee = 368,102 km
02	23:37	Mars (mag. 1.1) 2.4°S of the Moon
04	10:00 *	Jupiter (mag. -2.0) 0.1°N of Mercury (mag. -1.1)
05	00:37	Uranus (mag. 5.8) 3.7°S of the Moon
05	14:25	Jupiter (mag. -2.0) 4.7°S of the Moon
05	18:28	Mercury (mag. -1.2) 4.7°S of the Moon
06	04:21	Aldebaran 9.9°S of the Moon
06	12:38	New Moon
09	08:00	Pollux 1.7°N of the Moon
12	03:41	Regulus 3.3°S of the Moon
14	05:18	First Quarter
14	13:25	Moon at apogee = 404,077 km
16	18:11	Spica 1.2°S of the Moon
20	11:11	Antares 0.3°S of the Moon
22	01:08	Full Moon
27	11:30	Moon at perigee = 369,286 km
27	15:00	Saturn (mag. 1.1) 0.1°S of the Moon An occultation will be visible from parts of eastern Australia.
28	08:56	Neptune (mag. 7.9) 0.3°S of the Moon
28	21:53	Last Quarter

These objects are close together for an extended period around this time.

June 3 • *The Moon and Mars are side-by-side in the east-northeast. Menkar is lower and farther east (as seen from Sydney).*

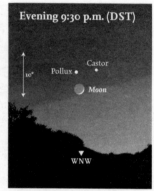

June 8 • *The Moon, Pollux and Castor form a right-angled triangle (as seen from central USA).*

J

June 9 • *The crescent Moon is close to Pollux and Castor. Procyon is higher and farther west (as seen from Sydney).*

June 16 • *The Moon and Spica are side by side (as seen from London).*

northwest · north · northeast

136

June – Looking North

For observers in the northern hemisphere, the time around the summer solstice (June 20) is frustrating for observing, because a form of twilight persists throughout the night. The sky remains so light that most faint stars and constellations are invisible unless conditions, such as the lack of light pollution, are particularly favourable. Even the highly distinctive set of seven stars forming the asterism of the 'Plough' (or 'Big Dipper') in **Ursa Major** may be difficult to see. The constellation has now swung round to the west, but the fainter stars of the constellation, lying even farther to the west, are harder to distinguish. The constellation of **Boötes** with bright **Arcturus** (α Boötis) is high overhead for northern observers, and best seen when facing south. The sprawling constellation of **Hercules** lies between **Lyra** and Boötes.

The four stars forming the 'head' of **Draco** lie east of the meridian, although only the brightest, **Eltanin** (γ Draconis) and, possibly **Rastaban** (β Draconis) are clearly visible. Farther north, in **Ursa Minor**, are **Polaris** itself and the two 'Guards', **Kochab** (β UMi) and **Pherkad** (γ UMi).

For observers around 50°N, the southern portions of the constellation of **Auriga** are partially lost on the northern horizon, although bright **Capella** (α Aurigae) should still be visible. Much of the neighbouring, fainter constellation of **Perseus** is difficult to make out. The two main stars of **Gemini**, **Castor** (α Geminorum) and **Pollux** (β Geminorum), the star closest to the ecliptic, and occasionally occulted by the Moon, are low on the northwestern horizon. Somewhat higher in the sky, and northeast of the meridian is **Cassiopeia** with **Cepheus** above it.

Two of the stars forming the angles of the distinctive Summer Triangle, **Deneb** (α Cygni) in **Cygnus** and **Vega** (α Lyrae) in Lyra are clearly visible as, for much of the night, is the third star, Altair (α Aquilae) in **Aquila**. Deneb lies in the Milky Way, at the beginning of the Great Dark Rift that runs down the constellation and where obscuring dust prevents us from seeing the dense star clouds of the Milky Way itself.

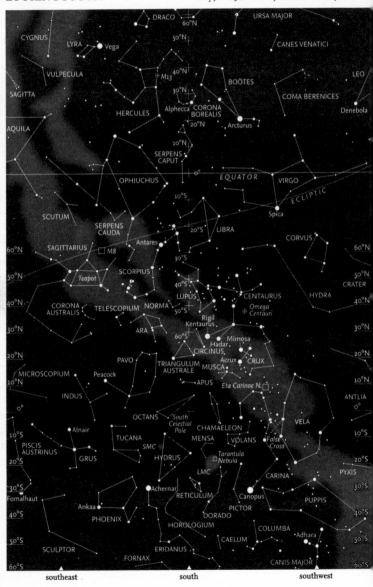

June – Looking South

The inconspicuous constellation of *Libra* is on the meridian, but the sky is really dominated by the striking constellation of *Scorpius*, slightly to the east. These two were once a single constellation, of course, until the 'claws' of the scorpion were formed into the constellation of The Balance. Although it is rather low for most northern observers, to those farther south, fiery red *Antares* (α Scorpii) is a major beacon and the whole constellation has a very significant presence in the winter skies. The constellation of *Virgo*, and bright *Spica* is now well to the west, along the ecliptic.

At this time of the year, the next constellation in the zodiac, *Sagittarius*, is becoming clearly seen, having risen in the east. The main body of the constellation forms the asterism known as 'The Teapot', but the constellation also has a long, curving chain of faint stars to the south, rather similar to the long line of stars below Scorpius, although curving in the opposite direction. This chain of stars partially encloses the small constellation of *Corona Australis*, although this constellation is not as well-formed as its northern counterpart (Corona Borealis). Rising in the east and following Sagittarius into the sky is the constellation of *Capricornus*. To the west of the 'sting' of Scorpius is the rather untidy tangle of stars that forms the constellation of *Lupus* and part of *Centaurus*.

The small constellation of *Crux* is now well south of the two brightest stars of Centaurus, *Rigil Kentaurus* and *Hadar*. The constellations of *Vela* and *Carina* and the *'False Cross'* are plunging towards the horizon. For observers at the latitude of Sydney in Australia, both *Canopus* (α Carinae) and *Achernar* (α Eridani) are so low that they are often invisible through absorption at the horizon.

The South Celestial Pole in *Octans* is surrounded by faint constellations: Octans itself, *Chamaeleon, Volans, Mensa, Hydrus, Tucana* and *Indus*. Only *Pavo*, to the northeast and *Grus*, to the southeast, are slightly more distinct. However, this area also includes the *Large Magellanic Cloud* (LMC), the largest satellite galaxy to our own. Most of the LMC lies in the constellation of *Dorado*, although some extends into the neighbouring constellation of Mensa.

Max Wolf

The famous German astronomer Maximilian Franz Joseph Cornelius (Max) Wolf was born at Heidelberg on 21 June 1863. He was largely instrumental in setting up the Heidelberg Observatory (the Landessternwarte Heidelberg-Königstuh) and became its director, a position which he held until his death in October 1932. He was a pioneer in astrophotography and carried out his work with the *Bruce double-astrograph*. He began by discovering a number of comets, but then used photographic means to detect minor planets, eventually discovering a total of 248.

He found a number of supernovae, and found that his photographic methods were particularly suitable for the study of the parallax of nearby stars. In 1919, he published a catalogue of the proper motions of over 1000 stars. One, in particular, a red dwarf known as *Wolf 359*, was later found to be one of the closest stars to the Solar System. Its distance is 7.9 light years, and it is of magnitude 13.54. The catalogue eventually contained over 1500 stars. Wolf was also instrumental in determining that the dark nebulae (William Herschel's 'holes in the sky') were actually clouds of dark dust that obscured more distant stars.

Max Wolf.

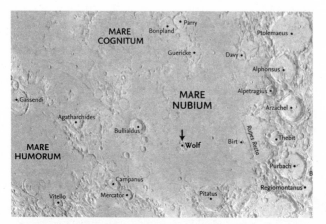

The lunar crater Wolf, approximately 25.7 km in diameter, located in the southern reaches of Mare Nubium, is named in his honour. The crater has been flooded by mare lava and the broken rim rises (in part) to about 0.7 km.

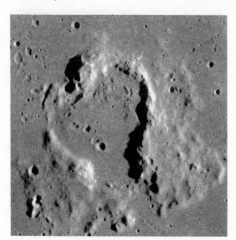

The impact crater Wolf, situated in the Mare Nubium, as imaged by the Lunar Reconnaissance Orbiter. North is at the top.

July

July – Introduction

Northern observers will continue to suffer from light nights during July, although there still remains the chance that they may be able to see noctilucent clouds during the first half of the month. (The noctilucent-cloud 'season' in the north lasts about 6 to 8 weeks, centred on the summer solstice.)

The Earth reaches aphelion, the farthest point from the Sun in its yearly orbit, at 05:06 UT on July 5. Its distance from the Sun is then 1.016725489 AU (which equals 152,099,894.38 km).

The first meteor shower that is active in July is that of the *α-Capricornids,* visible from both hemispheres. This a very long shower, beginning in early July and continuing until about mid-August. Unfortunately, the maximum rate is very low, around 5 meteors per hour, although there are occasional bright meteors. The peak occurs on July 30 when the Moon is a waning crescent. The parent body is a comet known as 169P/NEAT, one of the few comets discovered by NASA's Near-Earth Asteroid Tracking (NEAT) programme.

The second shower active in July is the *Southern δ-Aquariids,* also visible from both hemispheres. These begin in mid-month (on July 12, one day before First Quarter) and continue until August 23. Their maximum, like the α-Capricornids, is on July 30. Their hourly rate is higher than the α-Capricornids, and is about 25 meteors per hour. The parent body is uncertain, but believed to be Comet 96P/Macholz.

There is a minor southern shower, the *Piscis Austrinids,* which is a moderately long shower, beginning about July 15 and continuing until August 10. Like the α-Capricornids, the maximum hourly rate is very low, around 5 meteors per hour, with maximum on July 28, at Last Quarter. The parent body is currently unknown.

The most important shower that begins in July is the *Perseids.* This begins very unfavourably on July 17, when the Moon is waxing gibbous, but there is a strong maximum on August 12–13, so is more appropriately described next month.

The planets

Mercury is moving eastwards past the Sun, and comes to eastern elongation on July 22 at mag. 0.3. **Venus** is bright (mag. -3.9) in the evening sky. **Mars** (mag. 1.0 to 0.9) is in **Aries**, moving into **Taurus** at the end of the month. **Jupiter** (mag. -2.0 to -2.1) is moving slowly eastwards in Taurus. **Saturn** (mag. 1.0 to 0.9) is in **Aquarius** and begins retrograde motion on July 4. **Uranus** (mag. 5.8) is in **Taurus** and **Neptune** (mag. 7.9) remains in **Pisces**. Minor Planet **(1) Ceres** (mag. 7.3) comes to opposition on July 6 (see the maps on page 158).

The Japanese spacecraft **Hayabusa2** obtained sub-surface samples from the asteroid **Ryugu** on 11 July 2019.

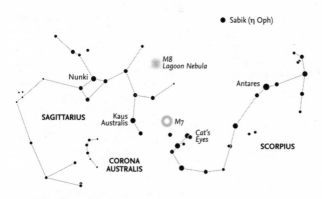

The sky looking south is dominated by the two zodiacal constellations of Scorpius and Sagittarius, on either side of the meridian, although they are low for observers at most northern latitudes.

Sunrise and sunset

City	Date	Sunrise	Sunset
Buenos Aires, Argentina			
	Jul. 01	11:01	20:53
	Jul. 31	10:48	21:12
Cape Town, South Africa			
	Jul. 01	05:52	15:48
	Jul. 31	05:40	16:06
London, UK			
	Jul. 01	03:47	20:22
	Jul. 31	04:22	19:52
Los Angeles, USA			
	Jul. 01	12:45	03:09
	Jul. 31	13:04	02:56
Nairobi, Kenya			
	Jul. 01	03:35	15:38
	Jul. 31	03:37	15:41
Sydney, Australia			
	Jul. 01	21:01	06:57
	Jul. 31	20:48	07:15
Tokyo, Japan			
	Jul. 01	19:29	10:01
	Jul. 31	19:48	09:47
Washington, DC, USA			
	Jul. 01	09:47	00:38
	Jul. 31	10:08	00:22
Wellington, New Zealand			
	Jul. 01	19:48	05:02
	Jul. 31	19:30	05:25

NB: the times given are in Universal Time (UT)

The Moon's phases and ages

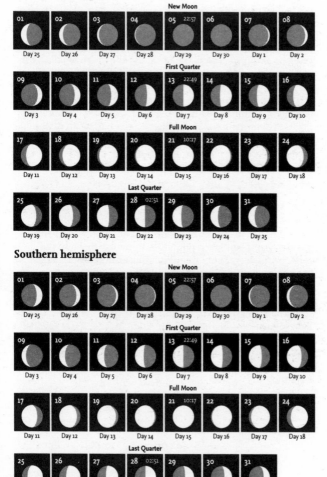

Northern hemisphere

New Moon

01 Day 25	02 Day 26	03 Day 27	04 Day 28	05 22:57 Day 29	06 Day 30	07 Day 1	08 Day 2

First Quarter

09 Day 3	10 Day 4	11 Day 5	12 Day 6	13 22:49 Day 7	14 Day 8	15 Day 9	16 Day 10

Full Moon

17 Day 11	18 Day 12	19 Day 13	20 Day 14	21 10:17 Day 15	22 Day 16	23 Day 17	24 Day 18

Last Quarter

25 Day 19	26 Day 20	27 Day 21	28 02:51 Day 22	29 Day 23	30 Day 24	31 Day 25

Southern hemisphere

New Moon

01 Day 25	02 Day 26	03 Day 27	04 Day 28	05 22:57 Day 29	06 Day 30	07 Day 1	08 Day 2

First Quarter

09 Day 3	10 Day 4	11 Day 5	12 Day 6	13 22:49 Day 7	14 Day 8	15 Day 9	16 Day 10

Full Moon

17 Day 11	18 Day 12	19 Day 13	20 Day 14	21 10:17 Day 15	22 Day 16	23 Day 17	24 Day 18

Last Quarter

25 Day 19	26 Day 20	27 Day 21	28 02:51 Day 22	29 Day 23	30 Day 24	31 Day 25

J

The Moon

The Moon in July

On July 1, the Moon is 4.1° north of **Mars** (mag. 1.0). The next day it is 4.0° north of **Uranus** (mag. 5.8). By July 3 it is 5.0° north of **Jupiter** (mag. -2.0) and, later, 9.9° north of **Aldebaran**. On July 6 the Moon is 3.9° north of **Venus** (mag. -3.9) and later 1.8° south of **Pollux** in **Gemini**. The next day the Moon is 3.2° north of **Mercury** (mag. -0.3). By July 9 it passes 3.0° north of **Regulus**, between it and **Algieba**. On July 14, one day after First Quarter, it is 0.9° north of **Spica**. On July 17 the Moon is 0.2° north of **Antares** in **Scorpius**. By July 24 the waning gibbous Moon is 0.4° north of **Saturn** (mag. 0.9) and the next day 0.6° north of **Neptune** (mag. 7.8). On July 29 one day after Last Quarter, the waning crescent Moon is 4.2° north of **Uranus** (mag. 5.8). By July 30 it passes 10.1° north of **Aldebaran**, 5.0° north of **Mars** (mag. 0.9) and 5.4° north of **Jupiter** (mag. -2.1).

Buck Moon

One of the names used in North America for the Full Moon for the month of July is 'Buck Moon'. This term derives from the fact that this is when new antlers grow on the heads of male (buck) deer. There are many other names for this Full Moon: among the Chippewa and Ojibwe of the Great Lakes area, following from the 'Strawberry Moon' of June, it is the 'Raspberry Moon'. Among the Arapaho of the Great Plains it is 'The Moon when the Hot Weather Begins' or 'The Moon when the Chokeberries Begin to Ripen'. Then, among the Omaha of the Central Plains it is 'the Moon when the buffalo bellow'. There are yet other terms from the Old English/Anglo-Saxon calendar, including the 'Thunder Moon', 'Wort Moon' and 'Hay Moon'.

Deep Impact

On 4 July 2005, the comet Tempel 1 (officially 9P/Tempel) was the subject of an experiment by NASA's ***Deep Impact*** spacecraft. The comet was struck by what was termed the 'Smart Impactor', a section with a mass of 372 kg, with a copper core. This copper 'Cratering Mass' had a mass of 113 kg. The larger, 601 kg section remained in orbit, to obtain images of the impact. The aim of the mission was to obtain details of the cometary composition.

The Smart Impactor was equipped with cameras and returned images until about 5 seconds before impact. These show craters on the surface of the comet. It is estimated that some 5 million kg of water and at least 10 million kg of dust were ejected by the impact. The dust cloud obscured the view of the impact crater from the main spacecraft. The surface material was found to be extremely fine: it was likened to talcum powder, rather than sand. Subsequent analysis of spectroscopic data revealed the presence of clays, carbonates, sodium, and crystalline silicates.

An image from the Smart Impactor (part of the Deep Impact *mission), showing craters on the surface of the comet Tempel 1.*

J

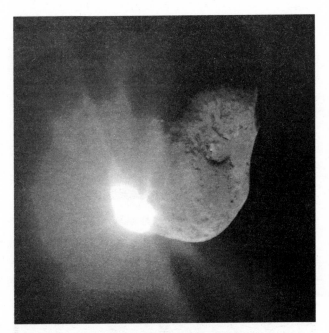

The moment of impact, as imaged by the main spacecraft.

Because of the problem of imaging the impact crater, NASA approved the New Exploration of Tempel 1 (or NExT) mission. This used the existing ***Stardust*** spacecraft, which had studied Comet Wild 2 in 2004. The ***Stardust*** mission obtained satisfactory images of the impact crater, which was determined to be about 150 metres.

The ***Deep Impact*** spacecraft was subsequently re-targetted (with the mission named EPOXI) to fly by comet Hartley 2 (103P/Hartley). It also observed Comet Garradd (C/2009 P1) from a considerable distance and ***Deep Impact*** observed Comet ISON in February 2013 until communications were lost in March 2013.

The image of Comet Hartley 2, obtained by the Deep Impact *spacecraft on 4 November 2010.*

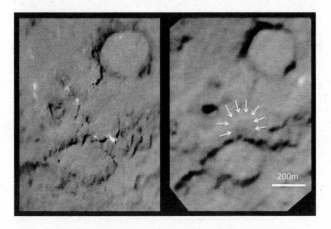

200m

Before and after images of the area around the impact crater, the right-hand image being obtained by the Stardust *spacecraft. The arrows outline the resulting crater.*

J

Calendar for July

01	18:26	Mars (mag. 1.0) 4.1°S of the Moon
02	10:07	Uranus (mag. 5.8) 4.0°S of the Moon
03–Aug.15		α-Capricornid meteor shower
03	08:28	Jupiter (mag. -2.0) 5.0°S of the Moon
03	12:01	Aldebaran 9.9°S of the Moon
05	05:06	Earth at aphelion
		(152,099,894.38 km = 1.016725 AU)
05	22:57	New Moon
06	00:04	Dwarf planet (1) Ceres at opposition (mag. 7.3)
06	15:04	Venus (mag. -3.9) 3.9°S of the Moon
06	16:29	Pollux 1.8°N of the Moon
07	18:32	Mercury (mag. -0.3) 3.2°S of the Moon
09	12:00	Regulus 3.0°S of the Moon
12–Aug.23		Southern δ-Aquariid meteor shower
12	08:11	Moon at apogee = 404,362 km
13	22:49	First Quarter
14	02:31	Spica 0.9°S of the Moon
15–Aug.10		Piscis Austrinid meteor shower
17–Aug.24		Perseid meteor shower
17	20:16	Antares 0.2°S of the Moon
21	10:17	Full Moon
22	06:39	Mercury at greatest elongation
		(26.9°E, mag. 0.3)
24	05:41	Moon at perigee = 364,917 km
24	20:46	Saturn (mag. 0.9) 0.4°S of the Moon
		An occultation of Saturn is visible from
		East Africa
25	14:54	Neptune (mag. 7.8) 0.6°S of the Moon
28		Piscis Austrinid meteor shower maximum
28	02:51	Last Quarter
29	17:30	Uranus (mag. 5.8) 4.2°S of the Moon
30		α-Capricornid meteor shower maximum
30		Southern δ-Aquariid meteor shower
		maximum
30	10:37	Mars (mag. 0.9) 5.0°S of the Moon
30	17:59	Aldebaran 10.1°S of the Moon
30	23:53	Jupiter (mag. -2.1) 5.4°S of the Moon

July 1–3 • *Before sunrise, the waning Moon passes Mars, the Pleiades, Jupiter and Aldebaran (as seen from central USA).*

July 7–9 • *The narrow crescent Moon passes Mercury, Regulus and Algieba (as seen from central USA).*

July 13 • *The First Quarter Moon is close to Spica, in the southwestern sky (as seen from central USA).*

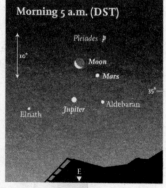

July 30 • *A nice gathering of the Moon, Jupiter, Mars, Aldebaran, the Pleiades and Elnath, shortly before sunrise (as seen from central USA).*

J

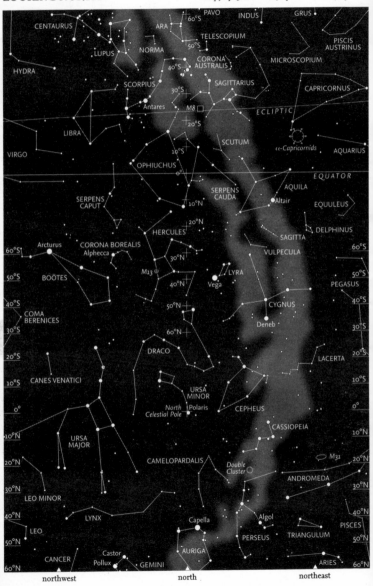

July – Looking North

The brilliant star **Vega** (α Lyrae) is now shining high overhead and the constellations of **Hercules** and **Lyra** are on opposite sides of the meridian, not far from the zenith for observers at 40°N, while it is the head of **Draco** that is near the zenith for observers slightly farther north at 50°N. Beyond Hercules, in the northwestern sky, are the constellations of **Corona Borealis** and **Boötes**, the latter with brilliant, orange-tinted **Arcturus**.

The stars of the Milky Way are now running more-or-less 'vertically', from north to south on the eastern side of the meridian. The constellation of **Cygnus** is 'upside down', high in the sky. For observers at the equator, it is the giant constellation of **Ophiuchus** that is at the zenith, with the two parts of **Serpens** (**Serpens Caput** to the west, and **Serpens Cauda** to the east, among the clouds of the Milky Way). Another star in the Summer Triangle with Vega, **Altair** (α Aquilae) in **Aquila** is similarly high in the sky.

Ursa Major is now clearly visible in the northwest, and on the opposite side of the meridian, the constellation of **Cepheus**, with its base in the Milky Way, is at a slightly greater altitude.

NASA's **Viking 1** spacecraft was the first successful Martian lander, touching down in western Chryse Planitia on 20 July 1976. It operated until 11 November 1982.

Cassiopeia, the other constellation that, like Ursa Major, is the key to finding one's way around the northern circumpolar constellations, lies in the Milky Way on the opposite side of the North Celestial Pole and **Polaris** in **Ursa Minor**. The faint constellations of **Camelopardalis** and **Lynx** lie to the west and slightly farther south. The chain of faint stars forming Lynx runs 'horizontally' below the outflung stars of Ursa Major. Below Cassiopeia on the other side of the meridian, **Perseus**, with the famous variable star, **Algol**, is beginning to climb higher in the sky and observers at mid-northern latitudes will find that they can now more clearly see **Capella** (α Aurigae) and the northernmost portion of **Auriga**. Observers in the far north (around latitude 60°N) may even occasionally glimpse **Castor** and **Pollux** in **Gemini** peeping above the northern horizon.

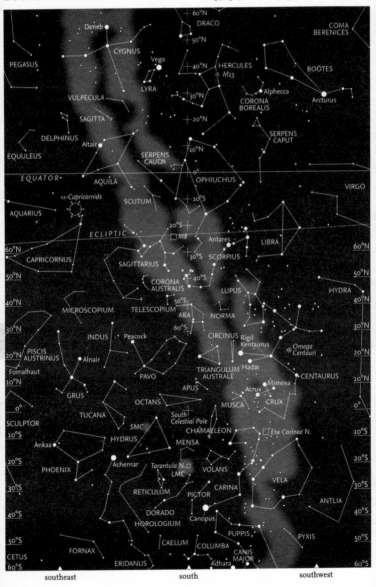

July – Looking South

The sky looking south is dominated by the two zodiacal constellations of **Scorpius** and **Sagittarius**, which lie on the ecliptic on either side of the meridian, although they are low for observers at most northern latitudes. Bright, red **Antares** (α Scorpii) is very conspicuous, even when it is low in the sky. Not for nothing has it earned the name that means 'Rival of Mars'. The roughly triangular shape of **Capricornus**, the next zodiacal constellation, lies east of Sagittarius and is now clearly visible. Below Scorpius, in the Milky Way, are the small, and often ignored, constellations of **Norma** (which is little more than three stars and easily overlooked) and **Ara**, the latter with a more distinctive shape and brighter stars. The stars of **Lupus** and the outlying stars of **Centaurus** (including the great globular cluster known as **Omega Centauri**) lie farther west.

High above, the constellation of **Ophiuchus** lies at the zenith for observers on the equator, with **Aquila** and bright **Altair** (α Aquilae) to the east. Northwest of Ophiuchus is **Boötes**, the principal star of which is **Arcturus**, which most people see as having an orange tint. Between Arcturus and the meridian is the circlet of stars that forms the constellation of **Corona Borealis**.

South of Scorpius is **Triangulum Australe**, to the east of the two bright stars of Centaurus, **Rigil Kentaurus** and **Hadar**, and lying within the star clouds of the Milky Way. At about the same altitude is the constellation of **Pavo** and, beyond it, **Indus**. Still farther east lies the elongated and somewhat distorted cross-shape that is the constellation of **Grus**. **Piscis Austrinus**, which lies south of Capricornus and Aquarius, has a single bright star, **Fomalhaut** (α Piscis Austrini). This ancient constellation (it was mentioned by Ptolemy, in the second century AD) has now risen in the east.

Crux and the adjoining constellation of **Musca** are now low on the horizon for observers at the equator, and the **Small Magellanic Cloud** (SMC) in **Hydrus** is actually below it. For observers farther south, **Achernar** (α Eridani) is now clearly visible, as is the constellation of **Phoenix** to the east. Only observers in the extreme south, however, will be able to see the whole of **Vela** and **Carina** as well as brilliant **Canopus** (α Carinae).

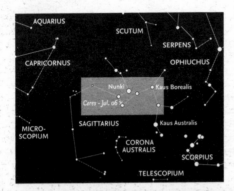

Minor planet (1) Ceres comes to opposition on July 6. The lighter grey box is shown in more detail below. The white cross marks the position of Juno at the day of its opposition.

The path of the minor planet (1) Ceres (mag. 8.7) around its opposition on July 6. On that day Ceres will be very close to the line connecting Ascella and τ Sgr. Background stars are shown down to magnitude 8.5.

Following its successful return of samples from minor planet (162173) Ryugu in December 2020, spaceprobe Hayabusa2 has been targetted to fly by the body known as (98943) 2001 CC21 in July 2026 and rendezvous with another, (1998) KY26 in July 2031.

August

August – Introduction

Meteors

August sees one of the finest meteor showers of the year. These are the famous **Perseids**, which are well-known, even to the general public, partly because they are visible during the warm nights of summer. In some Catholic countries they are known as the 'Tears of Saint Lawrence' because they are visible on August 10, the date of his martyrdom. They are said to represent sparks from the fire on which he was burnt. The Perseids are a very long shower, generally beginning about July 17 and continuing until around August 24, with a maximum on August 12–13, when the rate may reach as high as 100 meteors per hour (and on rare occasions, even higher). In 2024, maximum is when the Moon is around First Quarter, so conditions are not particularly good. The radiant lies in the northern portion of Perseus, towards the border with Camelopardus.

The Perseids are debris from Comet 109P/ Swift-Tuttle (the Great Comet of 1862) and there is a minor concentration that may give an early minor peak the day before the nominal maximum. Perseid meteors are fast and many of the brighter ones leave persistent trains. Some of the meteors may also produce bright fireballs.

There are three other meteor showers, all of which reach their peak maxima in July, but still exhibit activity that persists into August. These are the **α-Capricornids**, which may be active until 15 August (three days after First Quarter), the **Southern δ-Aquariids** (lasting until 23 August, three days before Last Quarter) and a southern shower, the **Piscis Austrinids**, which may persist until 10 August (two days before First Quarter). All these three showers have been described more fully on page 144.

On 18 August 1976 the Soviet lunar probe **Luna 24** landed on the Mare Crisium. After obtaining a surface sample, it lifted off from the Moon the following day, and returned the sample to Earth on 22 August 1976.

The planets

Mercury rapidly moves west and declines from mag. 0.9 to 2.2 before brightening at the end of the month to mag. 0.8. **Venus** is close by early in the month. **Mars** (mag. 0.8) is moving east in **Taurus**. **Jupiter** is also moving slowly east in **Taurus**. **Saturn** (mag. 0.7) continues retrograde motion throughout August. **Uranus** (mag. 5.8) is still in **Taurus** and **Neptune** (mag. 7.8) remains in **Pisces**. Minor planet (7) **Iris** (mag. 8.1) is at opposition on August 6.

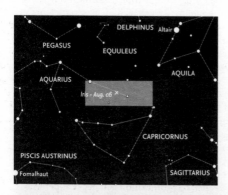

Minor planet (7) Iris is at opposition on August 6. The lighter grey box is shown in more detail below. The white cross marks the position of Iris at the day of its opposition.

A

The path of the minor planet (7) Iris (mag. 8.1) around its opposition on August 6. Background stars are shown down to magnitude 9.0.

Sunrise and sunset

City	Date	Sunrise	Sunset
Buenos Aires, Argentina			
	Aug. 01	10:47	21:13
	Aug. 31	10:14	21:34
Cape Town, South Africa			
	Aug. 01	05:39	16:06
	Aug. 31	05:06	16:27
London, UK			
	Aug. 01	04:24	19:50
	Aug. 31	05:11	18:50
Los Angeles, USA			
	Aug. 01	13:04	02:55
	Aug. 31	13:26	02:22
Nairobi, Kenya			
	Aug. 01	03:37	15:41
	Aug. 31	03:31	15:36
Sydney, Australia			
	Aug. 01	20:47	07:15
	Aug. 31	20:14	07:36
Tokyo, Japan			
	Aug. 01	19:49	09:46
	Aug. 31	20:12	09:11
Washington, DC, USA			
	Aug. 01	10:09	00:21
	Aug. 31	10:36	23:40
Wellington, New Zealand			
	Aug. 01	19:28	05:26
	Aug. 31	18:47	05:55

NB: *the times given are in Universal Time (UT)*

The Moon's phases and ages

Northern hemisphere

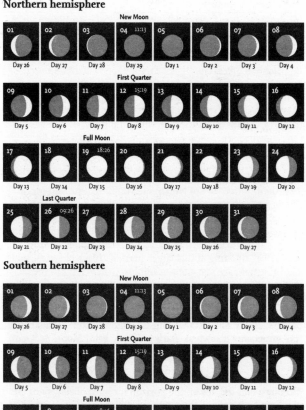

New Moon

			04 11:13				
01	02	03	04	05	06	07	08
Day 26	Day 27	Day 28	Day 29	Day 1	Day 2	Day 3	Day 4

First Quarter

			12 15:19				
09	10	11	12	13	14	15	16
Day 5	Day 6	Day 7	Day 8	Day 9	Day 10	Day 11	Day 12

Full Moon

		19 18:26					
17	18	19	20	21	22	23	24
Day 13	Day 14	Day 15	Day 16	Day 17	Day 18	Day 19	Day 20

Last Quarter

	26 09:26					
25	26	27	28	29	30	31
Day 21	Day 22	Day 23	Day 24	Day 25	Day 26	Day 27

Southern hemisphere

New Moon

			04 11:13				
01	02	03	04	05	06	07	08
Day 26	Day 27	Day 28	Day 29	Day 1	Day 2	Day 3	Day 4

First Quarter

			12 15:19				
09	10	11	12	13	14	15	16
Day 5	Day 6	Day 7	Day 8	Day 9	Day 10	Day 11	Day 12

Full Moon

		19 18:26					
17	18	19	20	21	22	23	24
Day 13	Day 14	Day 15	Day 16	Day 17	Day 18	Day 19	Day 20

Last Quarter

	26 09:26					
25	26	27	28	29	30	31
Day 21	Day 22	Day 23	Day 24	Day 25	Day 26	Day 27

A

The Moon

The Moon in August

On August 2, the Moon is 1.8° south of **Pollux**. By August 5 it is 2.9° north of **Regulus** in **Leo**. Later the same day it is 1.7° north of **Venus** (mag. -3.8). On August 6 the Moon is 7.5° north of **Mercury** and close to **Venus**. By August 10 the waxing crescent Moon is 0.7° north of **Spica**. On August 14, the Moon occults **Antares**, but this is visible only over the Pacific Ocean. By August 21, the Moon is 0.5° north of **Saturn** (mag. 0.7) and later that day is 0.7° north of **Neptune** (at mag. 7.8). By August 26 the Moon is 4.4° north of **Uranus** (mag. 5.8) and, later, 10.3° north of **Aldebaran**. On August 28, the Moon passes 5.3° north of **Mars** and on August 30 1.7° south of **Pollux**.

Sturgeon Moon

The Algonquin tribes of North America called the Full Moon of August the 'Sturgeon Moon' because of the numerous examples that they captured in the lakes where they fished. To the Cree of the Canadian Northern Plains it was the 'Moon when young ducks begin to fly'. Many tribal names refer to the berries ripening at this time: 'berries', 'black cherries', and 'chokeberries', are just a few examples. Others refer to the fact that corn is ripening, such as among the Ponca of the Southern Plains where it was the 'Corn is in the Silk Moon'. To the Haida of Alaska it was the 'Moon for cedar bark for hats and baskets'. In the ancient Old English/Anglo-Saxon calendar it was sometimes known as 'Barley Moon', 'Fruit Moon' or 'Grain Moon'.

Sagittarius and Scorpius

Although they are low on the horizon for most observers in the northern hemisphere, these two constellations contain much of interest. Sagittarius, in particular, has many fascinating objects. It contains the **Teapot** asterism, which is extremely distinctive. Because the constellation harbours the centre of the Milky Way galaxy, it has no fewer than fifteen Messier objects and many additional globular clusters and emission nebulae. Other objects are undoubtedly hidden by the vast amounts of interstellar dust that lie in this region. There are a few 'windows' in this dust, however, that allow more distant objects to be seen. One such object is **NGC 6822**, 'Barnard's Galaxy', an irregular galaxy in the far northwest of the constellation, which lies outside our own stellar system and is part of the Local Group of galaxies. Much closer to the centre of the galaxy is the area known as 'Baade's Window' where there is little obscuring dust, enabling two distant globular clusters to be seen.

Almost due north of the tip of the Teapot's 'Spout' (the star **Alnasl** or γ **Sgr**), is the galactic centre, and the location of the (invisible) supermassive black hole known as 'Sagittarius A*' (abbreviated as **Sgr A***). The mass of this black hole has been determined from the orbits of several stars around the centre. The current estimate is that it probably contains more than 4 million solar masses. One star, known as S4714, has been determined to be the star that comes closest to the central black hole (at 'perinigricon') at a distance of about 12.6 AU, less than Uranus' distance from the Sun. (Just as we have 'perigee' for the closest approach to Earth, and 'perihelion' for the closest

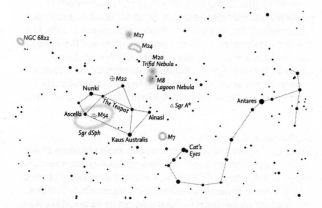

The constellation of Sagittarius not only contains the galactic centre (Sgr A*) but also many interesting objects. The satellite galaxy, the Sagittarius Dwarf Spheroidal Galaxy (Sgr dSph), is hidden by clouds of dust.

A highly detailed image of the Lagoon Nebula (M8) recently obtained by the VLT Survey Telescope (VST) at Paranal Observatory in Chile.

approach to the Sun, we now have 'perinigricon' for the closest approach to a black hole.)

Slightly southwest of the Galactic Centre is **M8**, the 'Lagoon Nebula', which is just visible to the naked eye under good conditions. This is a giant star-forming interstellar cloud, primarily a giant emission nebula, illuminated by the young, hot stars within it. Slightly farther north again is **M20**, the 'Trifid Nebula', which is actually a rare combination of three different forms of nebulosity: an emission nebula (pinkish), where the radiation from young stars is causing the surrounding hydrogen gas to glow; a reflection nebula (bluish in tint); and a dark dust nebula that is causing the dark 'lanes', dividing the object into three lobes. Very slightly to the south of M20 and farther east is **M22**, a bright (mag. 5) globular cluster, visible to the naked eye. One of Messier's objects, **M24**, northeast of Kaus Borealis (λ Sgr) is not a true open cluster, but consists of a dense concentration of stars, known as the Small Sagittarius Star Cloud. (The Large Sagittarius Star Cloud is the dense portion of the galaxy's central bulge, visible below – southwest of – the Great Dark Rift.)

Much farther north, and near the point that divides the constellations of Serpens and Scutum from Sagittarius, is **M17**, which is known both as the 'Omega Nebula' and also as the 'Swan Nebula'. This is a gaseous nebula, best seen in binoculars or a telescope.

The globular cluster **M54** lies southeast of the star **Ascella** (ζ Sgr) that is the base of the 'Handle' of the 'Teapot'. It is actually part of the Sagittarius Dwarf Spheroidal Galaxy (**Sgr dSph**) which was discovered much later (1994) than M54 itself, and which is now known to be one of the Galaxy's satellite galaxies.

Just over the southwestern border of the constellation and into the constellation of Scorpius, slightly farther south than the star **Kaus Australis** (ε Sgr), the base of the 'Spout' of the 'Teapot', is **M7** (sometimes known as the Ptolemy Cluster), a very large, bright open cluster, that is mag. 3.5 and thus readily visible to the naked eye, but best seen with low magnification in binoculars.

A

An all-sky image of the Galaxy and the two Magellanic Clouds obtained by the Gaia spacecraft. The satellite galaxy, the Sagittarius Dwarf Spheroidal Galaxy (Sgr dSph), is a barely-visible streak below the galactic centre.

Calendar for August

02	23:35	Pollux 1.8°N of the Moon
04	11:13	New Moon
05	19:34	Regulus 2.9°S of the Moon
05	22:03	Venus (mag. -3.8) 1.7°S of the Moon
06	00:02	Mercury (mag. 1.6) 7.5°S of the Moon
06	15:00 *	Venus (mag. -3.8) 5.9°N of Mercury (mag. 1.6)
06	19:40	Minor planet (7) Iris at opposition (mag. 8.1)
09	01:31	Moon at apogee = 405,297 km
10	10:17	Spica 0.7°S of the Moon
12–13		Perseid meteor shower maximum
12	15:19	First Quarter
14	05:17	Antares 0.0°N of the Moon. There is an occultation of Antares, visible only over the Pacific Ocean
19	18:26	Full Moon
21	03:02	Saturn (mag. 0.7) 0.5°N of the Moon
21	05:02	Moon at perigee = 360,196 km
21	22:21	Neptune (mag. 7.8) 0.7°S of the Moon
26	00:01	Uranus (mag. 5.7) 4.4°S of the Moon
26	09:26	Last Quarter
26	23:29	Aldebaran 10.3°S of the Moon
27	12:43	Jupiter (mag. -2.3) 5.7°S of the Moon
28–Sep.05		α-Aurigid meteor shower
28	00:22	Mars (mag. 0.8) 5.3°S of the Moon
30	05:25	Pollux 1.7°N of the Moon

These objects are close together for an extended period around this time.

August 6 · The Moon is close to Venus, Regulus and Mercury. Algieba and Denebola are close by (as seen from Sydney).

August 10 · Due west, the Moon and Spica are close together (as seen from Sydney).

August 26–28 · After Last Quarter, the Moon moves along the Pleiades, Aldebaran, Jupiter, Mars and Elnath (as seen from London).

August 30 · The Moon lines up with Pollux and Castor (as seen from London).

A

August – Looking North

The Milky Way is now running up the eastern side of the sky, where **Cassiopeia** may be found within it, high in the northeast. The constellation of **Cepheus**, with the 'base' of the constellation, which is shaped like the gable-end of a house, lies within the edge of the Milky Way, and the whole constellation is 'upside-down' slightly farther north than Cassiopeia. For observers south of the equator, these constellations are close to the horizon and **Ursa Minor** is on the horizon.

For most mid-latitude northern observers, **Perseus** is now clearly visible, as is the northern portion of **Auriga** with brilliant **Capella** (α Aurigae). Perseus straddles a narrow portion of the Milky Way and although not particularly distinct, a significant section of the Milky Way actually runs through Auriga in the direction of **Gemini**. Auriga contains numerous open clusters and a few emission nebulae, but is without globular clusters. Beyond Perseus in the east lies **Andromeda** and the small constellation of **Triangulum**. Slightly farther north than Perseus and on the other side of the meridian is **Ursa Major**, and for most observers even the stars in the southern, extended portion are visible.

For observers at about 40°N, both **Lyra** and **Cygnus**, with their brilliant principal stars, **Vega** and **Deneb**, respectively, are high overhead, near the zenith. The third star marking the other apex of the Summer Triangle, **Altair** in **Aquila**, is much farther to the south, and all three stars are best seen when looking south. The constellation of **Hercules** lies farther west of Lyra and Cygnus, with most of the constellation of **Pegasus** to the east. The tiny, but highly distinctive constellation of **Delphinus** lies between the Milky Way and the outlying stars of Pegasus. Again, these areas are best seen when looking south.

A

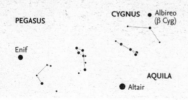

There are four small constellations between Cygnus, Pegasus and Aquila. They are, from left to right: Equuleus, Delphinus, Sagitta and Vulpecula.

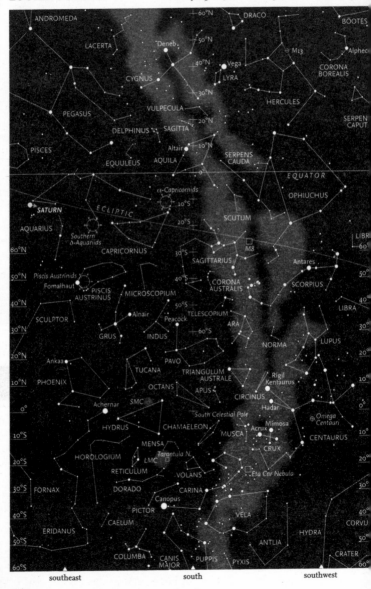

southeast   south   southwest

172

August – Looking South

The star clouds of the Milky Way dominate the sky at this time of the year. The three stars forming the Summer Triangle, **Deneb** (α Cygni), **Vega** (α Lyrae) and **Altair** (α Aquilae) are clearly visible, as is the Great Dark Rift, running south from near Deneb. The Great Dark Rift marks the location of dense clouds of dust that obscure the light from the many stars in the main plane of the Galaxy. Both **Cygnus** and **Aquila** represent birds, and in modern charts these are shown flying 'down' the Milky Way, towards the south. Older charts tended to show Aquila as flying 'across' the Milky Way, in the direction of **Aquarius**, in the east.

For observers at northern latitudes, the constellations of **Hercules** and **Pegasus** are visible, one on each side of the meridian, together with **Delphinus** on the east. Between Cygnus and Aquila, in the Milky Way, lie the two small constellations of **Vulpecula** and **Sagitta**.

The two zodiacal constellations of **Scorpius** and **Sagittarius** are still clearly visible, although becoming rather low for observers at mid-northern latitudes, with even the bright red supergiant star of **Antares** in Scorpius becoming difficult to see. (For observers at 50°N, it is skimming the horizon.) North of Scorpius is the large constellation of **Ophiuchus**, with part of **Serpens**, **Serpens Cauda**, lying in front of the Great Dark Rift.

Capricornus is clearly seen, as is **Aquarius**, the next constellation along the ecliptic. For observers farther south, the curl of stars forming **Corona Australis** lies south of Sagittarius and, farther east, **Piscis Austrinus** with its single bright star, **Fomalhaut** (α Piscis Austrini) is at about the same altitude, with the chain of faint stars that curves south from Sagittarius and the faint constellation of **Microscopium** in between the two. South of Piscis Austrinus is the constellation of **Grus**, with the undistinguished constellation of **Indus** between it and **Pavo**, which is on the meridian.

Farther south, **Lupus**, **Centaurus** and **Crux** are descending towards the horizon, while for observers at 30°S, **Vela** has disappeared, as has most of **Carina**, including brilliant **Canopus**.

A

September

September – Introduction

The (northern) autumnal equinox occurs on September 22 in 2024, when the Sun moves south of the equator in the western side of the constellation of **Virgo**. This position is sometimes known as the **First Point of Libra**, which is where it lay in former times. However, precession has now shifted this point over the border and into the constellation of **Virgo** (see the map below).

Meteors

After the major Perseid shower in August, there is very little meteor shower activity in September. One minor northern shower, called the **α-Aurigids**, tends to have two peaks of activity. The principal peak occurs on September 1. In 2024, the Moon is a thin waning crescent, so conditions are very good, apart from the fact that the maximum rate is only 5 or 6 meteors per hour, although the meteors are bright and relatively easy to photograph. Activity from this shower may even extend into October. The **Southern Taurid** shower begins this month (on September 10) and, although rates are similarly low, it often produces very bright fireballs. This is a very long shower, lasting until about November 20, with a maximum in 2024 on October 10–11. This year, activity begins when the moon is just before First Quarter (Day 7), and peaks at about the same age (about Day 8 or 9 of the lunation). As a slight compensation for the lack of shower activity, however, the number of sporadic meteors visible in the month of September reaches its highest rate at any time during the year.

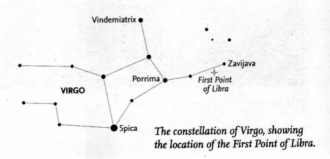

The constellation of Virgo, showing the location of the First Point of Libra.

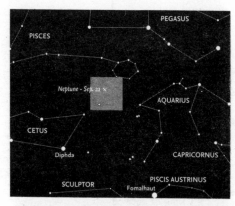

A finder chart for the position of Neptune at its opposition on September 21. The grey area is shown in more detail below. The white cross marks the position of Neptune at the day of its opposition.

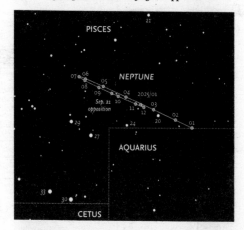

The path of Neptune in 2024. Stars down to magnitude 8.5 are shown.

S

The Planets

Mercury (mag. 0.3) reaches greatest western elongation on
September 5. *Venus* (mag. -3.8) moves from *Virgo* into *Libra*, well
east of the Sun in the evening sky. *Mars* (mag. 0.6) moves from
Taurus into *Gemini*. *Jupiter* is moving slowly east in Taurus.
Saturn is still retrograding in *Aquarius* and comes to opposition
on September 8 (see the chart on page 190). *Uranus* (mag. 5.7)
begins retrograde motion on September 15 in Aquarius. *Neptune*
(mag. 7.8) is in *Pisces* and comes to opposition on September 21,
close to the point of the vernal equinox (see the chart on page 177).

Viewing Earth satellites

There are now so many objects orbiting the Earth that if you
watch the sky shortly after sunset or before dawn, you are
almost certain to see a tiny spot of light crossing the sky. Why
'shortly after sunset or before dawn'? Well, the satellite itself
must be illuminated by the Sun, while you are in darkness.
Satellites often disappear as they pass into Earth's shadow.
Even the massive constellations of satellites that are now being
launched and which will seriously jeopardise astronomy in

*A sequence of
images, taken with
a specially equipped
telescope of the*
International
Space Station *as
its path took it in
front of the Sun.*

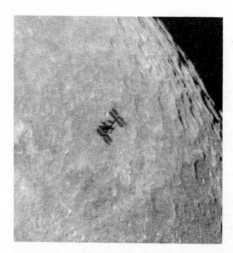

The International Space Station, *photographed in front of Mare Serenitatis on the Moon. The large crater to the right of the ISS is Posidonius.*

the evening and morning are likely to suffer in the same way. Even those satellites will disappear as they are overtaken by the shadow. Because of their physical composition, sometimes with large flat panels that are highly reflective, satellites may not reflect sunlight evenly in all directions, and many vary in brightness, or even 'flare' as they pass across the sky. The Iridium series of communication satellites are particularly noted for their great variations in brightness.

There are numerous websites that offer details of when satellites may be seen from your location and some are given later under 'Further Information'. With some you may request emails to be sent to you with details of when specific satellites may be visible from your position on Earth.

The largest artificial object in orbit around the Earth is, of course, the ***International Space Station (ISS)***. The ISS orbits at an average altitude of 400 kilometres and completes one orbit in about 93 minutes. Once again, you can obtain predictions of where it may be seen from your location. It generally appears as a very bright object, but may occasionally be seen as a dark body passing in front of the Moon.

S

Sunrise and sunset

City	Date	Sunrise	Sunset
Buenos Aires, Argentina			
	Sep. 01	10:13	21:35
	Sep. 30	09:32	21:56
Cape Town, South Africa			
	Sep. 01	05:05	16:28
	Sep. 30	04:25	16:48
London, UK			
	Sep. 01	05:13	18:48
	Sep. 30	05:59	17:42
Los Angeles, USA			
	Sep. 01	13:27	02:20
	Sep. 30	13:47	01:40
Nairobi, Kenya			
	Sep. 01	03:30	15:35
	Sep. 30	03:19	15:26
Sydney, Australia			
	Sep. 01	20:13	07:37
	Sep. 30	19:33	07:57
Tokyo, Japan			
	Sep. 01	20:13	09:09
	Sep. 30	20:35	08:27
Washington, DC, USA			
	Sep. 01	10:37	23:39
	Sep. 30	11:03	22:53
Wellington, New Zealand			
	Sep. 01	18:45	05:56
	Sep. 30	17:56	06:25

NB: the times given are in Universal Time (UT)

The Moon's phases and ages

Northern hemisphere

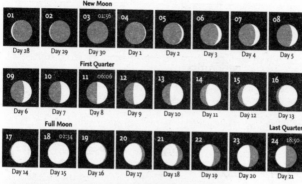
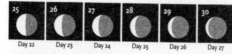

Southern hemisphere

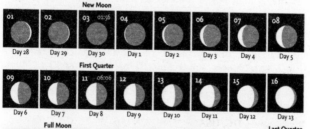

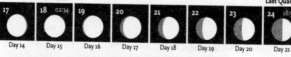
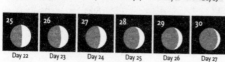

The Moon

The Moon in September
On September 5, the waxing crescent Moon (two days after New) is 1.2° south of **Venus** (which is bright at mag. -3.8). By September 17, one day before Full, the Moon is 0.3° north of **Saturn**. At Full Moon it is 0.7° north of **Neptune**, which will be difficult to see (at mag. 7.8) On September 22, it is 4.5° north of **Uranus** (which is mag. 5.8) and the next day, September 23, just before Last Quarter, it is 5.7° north of brilliant **Jupiter** (mag. -2.4). On September 25 it is 4.9° north of **Mars** (which is mag. 0.6).

Corn Moon
To the peoples of North America, September was particularly important because corn (maize) was ready for harvest. Many tribes had names for the Full Moon that referred to corn, such as 'Corn Maker Moon' among the Abenaki of northern Maine; 'Middle Moon between harvest and eating corn' to the Algonquin in the Northeast and Great Lakes area, although it was the 'Drying Grass Moon' for the Cheyenne of the Great Plains.

The Japanese spacecraft *Hayabusa2* dropped two surface rovers onto the asteroid *Ryugu* on 21 September 2018. A further rover was deployed on 3 October 2018.

In Europe, the September Full Moon was generally called the 'Harvest Moon', and technically this was the first Full Moon after the autumnal equinox (22 September in 2024). In general, this Full Moon comes in September, but approximately once every three years, Full Moon comes in October, when it is that Full Moon that is known as the 'Harvest Moon'. The term 'Harvest Moon' was the only name for the Full Moon that was determined by the equinox, rather than being specific to any particular month. Other names for this Full Moon, from the Old World, and specifically from the Old English/Anglo-Saxon calendar, were 'Full Corn Moon' – in this case 'corn' meaning wheat or barley – as well as 'Barley Moon'.

The surface of Pluto, as imaged by the New Horizons *spaceprobe. The large, apparently flat region is known as Sputnik Planitia.*

New Horizons

The *New Horizons* spaceprobe was launched in January 2006. The spacecraft was designed to make observations of Pluto during a fly-by on 14 July 2015, and also make observations of more distant objects in the Kuiper Belt beyond it. (These objects – the Kuiper Belt Objects (KBO), which include Pluto – are strongly affected by the gravitational influence of Neptune.)

The *New Horizons* probe was able to obtain images of the surface of Pluto and of its large satellite, Charon. The observations confirmed the existence of a thin atmosphere and atmospheric hazes. Water ice has been detected on parts of the surface and details of the surface have now been confirmed to show convective cells of nitrogen ice.

After its fly-by of Pluto, the spaceprobe was directed to the KBO 486958, now known as Arrokoth.

An image of the Kuiper Belt Object, now known as Arrokoth, as obtained by the New Horizons *spaceprobe.*

S

Calendar for September

01		α-Aurigid meteor shower maximum
01	09:16	Mercury (mag. 0.5) 5.0°S of the Moon
03	01:56	New Moon
05	02:30	Mercury at greatest elongation (18.1°W, mag. -0.3)
05	10:16	Venus (mag. -3.8) 1.2°N of the Moon
05	14:54	Moon at apogee = 406,211 km
08	04:35	Saturn at opposition (mag. 0.6)
10–Nov.20		Southern Taurid meteor shower
11	06:06	First Quarter
17	10:22	Saturn (mag.0.6) 0.3°S of the Moon
18	02:34	Full Moon
18	02:45	Partial lunar eclipse
18	07:35	Neptune (mag. 7.8) 0.7°S of the Moon
18	13:22	Moon at perigee = 357,286 km
21	00:17	Neptune at opposition (mag. 7.8)
22	07:14	Uranus (mag. 5.7) 4.5°S of the Moon
22	12:44	September equinox
23	23:21	Jupiter (mag. -2.4) 5.7°S of the Moon
24	18:50	Last Quarter
25	11:49	Mars (mag. 0.6) 4.9°S of the Moon
29		Daylight Saving Time begins (New Zealand)

September 17 • *The Moon and Saturn are close together, with Fomalhaut higher and farther south (as seen from Sydney).*

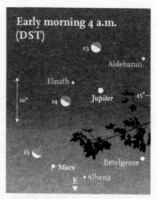

September 23–25 • *The Moon passes Aldebaran, Elnath and Mars, high in the eastern sky (as seen from central USA).*

September 24–27 • *The Last Quarter Moon moves from the north to the northeast, and passes Jupiter, Elnath, Alhena, Mars, Pollux and Castor (as seen from Sydney).*

S

LOOKING NORTH

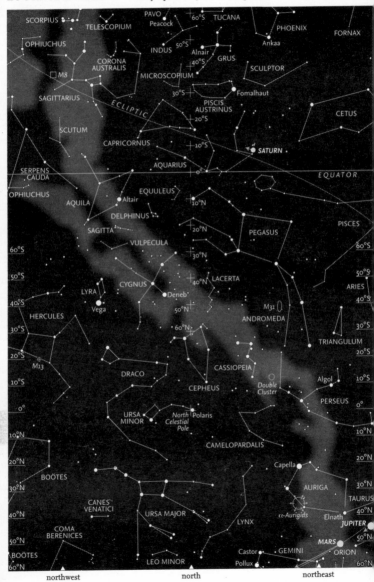

September – Looking North

For mid-northern observers, **Cassiopeia** is high overhead, and **Cepheus** is 'upside-down', apparently hanging from the Milky Way, high above the Pole. **Perseus** is high in the northeast, at about the same altitude as **Ursa Minor** and **Polaris**.

For those same observers, **Ursa Major** is now 'right way up', low in the south, although some of the southernmost stars are lost along or below the horizon. **Auriga**, with bright **Capella**, and the small triangle of the 'Kids' is clearly visible in the northeast. Higher in the sky, above Perseus, the whole of **Andromeda** is visible, together with the small constellation of **Triangulum** and, still higher, the Great Square of **Pegasus**, and beyond it, most of the zodiacal constellation of **Pisces**.

The clouds of stars forming the Milky Way are not particularly striking in Auriga and Perseus, but beyond Cassiopeia, and on towards **Cygnus**, they become much denser and easier to see.

Cygnus is high in the northwest and **Deneb**, its principal star, is close to the zenith for observers at 40–50° north, with **Lyra** and bright **Vega**, slightly farther west. Farther towards the south, most of **Hercules** is clearly visible, with the 'Keystone' and the globular cluster M13. The head of **Draco** lies between Hercules and Cepheus and the whole of that constellation is easily seen as it curls around Ursa Minor and the North Celestial Pole.

Observers in the far north may be able to detect **Castor** (α Geminorum) skimming the northern horizon, although brighter **Pollux** (β Geminorum) will be too low to be detected until later in the night.

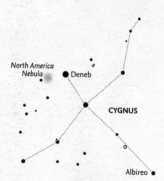

The constellation of Cygnus. Deneb (α Cyg) is one of the stars forming the Summer Triangle. The other two are Vega (α Lyr) and Altair (α Aql).

S

September – Looking South

The three stars forming the apices of the (northern) Summer Triangle, **Deneb** (α Cygni), **Vega** (α Lyrae) and **Altair** (α Aquilae) are prominent in the southwest. The Great Dark Rift is clearly visible, starting near Deneb and running down the centre of the Milky Way towards the centre of the Galaxy in **Sagittarius**, and beyond into **Scorpius**, only petering out in **Centaurus** and **Crux**.

Most of the constellation of **Capricornus** lies just west of the meridian, with **Aquarius** just slightly farther north on the eastern side. The next zodiacal constellation, **Pisces**, is clearly seen, including the prominent asterism, known as the 'Circlet'. Much of the constellation of **Cetus** is visible south of it. South of Aquarius is **Piscis Austrinus**, with its single bright star, **Fomalhaut**, in an otherwise fairly barren area of sky. Still slightly farther south is the undistinguished constellation of **Sculptor**, and, closer to the meridian, the line of stars that forms part of **Grus**.

For observers south of the equator, Sagittarius is high in the west. Below it is the curl of stars that is **Corona Australis**. Below (south of) Sagittarius is the tail and 'sting' of Scorpius. Still farther south is the constellation of **Lupus** and the scattered stars of **Centaurus**, with **Rigil Kentaurus** and **Hadar** (α and β Centauri, respectively). Between the 'sting' of Scorpius and those two bright stars are the small constellations of **Ara** and **Triangulum Australe**. Next comes Crux itself. If the long axis of the cross is extended right across the sky over the largely empty area of sky around the South Celestial Pole, it points in the general direction of **Achernar** (α Eridani). Between Achernar and Ara, west of the meridian, lies the constellation of **Pavo** with it sole bright star **Peacock** (α Pavonis). To the south, the brightest star of the southern hemisphere, **Canopus** (α Carinae) is hugging the horizon for observers at the latitude of Sydney in Australia.

Fomalhaut   PISCIS AUSTRINUS

For observers at 30°S, the constellation of Piscis Austrinus, with its single bright star, Fomalhaut, is almost overhead.

S

In 2024, Saturn (mag. 0.6) comes to opposition on September 8, in the constellation of Aquarius. The numbers relating to the positions of Saturn in the diagram above correspond to the months of the year.

On 5 September 1977, the **Voyager 1** spaceprobe was launched. It studied Jupiter, Saturn and Saturn's satellite, Titan. It became the first object to leave the heliosphere and enter interstellar space.

October

October – Introduction

There is one major meteor shower (the **Orionids**) and three minor showers active during October. The main shower is the Orionids which are fairly reliable in October. Like the May *η-Aquariid* shower, the Orionids are associated with Comet 1P/Halley. During this second pass through the stream of particles from the comet, slightly fewer meteors are seen than in May, but conditions are more favourable for northern observers. In both showers the meteors are very fast, and many leave persistent trains. Although the Orionid maximum is quoted as October 21–22 for 2024, in fact there is a very broad maximum, lasting about a week, roughly centred on that date, with hourly rates around 25. Occasionally, rates are higher (even reaching 50–70 per hour). In 2024, the Moon is approaching Last Quarter at the nominal maximum, so conditions are not particularly favourable.

The faint shower of the **Southern Taurids** (often with bright fireballs) peaks on October 10–11. The Southern Taurid maximum occurs around First Quarter, so conditions are not particularly favourable. Towards the end of the month (around October 20), another shower (the **Northern Taurids**) begins to show activity, which peaks early in November. The parent comet for both Taurid showers is Comet 2P/Encke, which has the shortest period (3.3 years) of any major comet. The meteors in both Taurid streams are relatively slow and bright.

A short, minor northern shower, the **Draconids**, begins on October 6 and peaks on October 8–9. Although the rate is about 10 meteors per hour, the whole shower occurs around First Quarter, so conditions are not very favourable in 2024.

The planets
Mercury (mag.-1.6) moves away from the Sun. **Venus** (mag.-3.9 to -4.0) is moving eastwards in Virgo. **Mars** (changing rapidly from mag.0.5 to mag.0.1) is in **Gemini**, moving towards **Cancer**. **Jupiter** (mag.-2.5 to -2.7) begins retrograde motion on October 9 and is in **Taurus**. **Saturn** (mag.0.7 to 0.8) is still retrograding slowly in **Aquarius**. **Uranus** (mag.5.6) is also retrograding in **Taurus**. **Neptune** (mag.7.8) remains in **Pisces**.

Olbers' Paradox

The German astronomer Heinrich Wilhelm Olbers (1758–1840) posed a paradox about the night sky. If the universe were static (unchanging), homogeneous (i.e., the same everywhere) and with a population of an infinite number of stars, then any line of sight must eventually encounter the surface of a star. So the universe should appear uniformly bright wherever one looked and thus be light at night. Obviously one (at least) of the three basic assumptions must be incorrect. The most widely accepted explanation is that the universe is expanding away – or rather, that spacetime is expanding – from its origin in the Big Bang. The expansion causes the redshift of light from any stars or galaxies towards longer wavelengths. Eventually the light becomes invisible to human eyes, hence the darkness at night. This may be taken as evidence that the universe is dynamic, i.e., changing.

The problem of an infinite number of stars in the universe and its significance for the darkness at night was commented upon by a Greek monk, Cosmas Indicopleustes, as long ago as the sixth century, but Olbers was the first to put the problem into a paradoxical form. The paradox is sometimes known as 'the dark night sky paradox'.

There are other theories that account for the darkness of the night sky, without having recourse to a Big Bang from which spacetime is expanding, causing the light from distant stars and galaxies to be shifted into the red; to longer and longer wavelengths, until it becomes invisible to human eyesight and our telescopes. One such theory is the 'Fractal Universe' in which stars occur in a fractal distribution where there are fewer stars, that is, the density of stars decreases as the volume considered increases. This would lead to a dark night sky, but does not rule out that the Universe came into being in a Big Bang.

O

Sunrise and sunset

City	Date	Sunrise	Sunset
Buenos Aires, Argentina			
	Oct. 01	09:30	21:56
	Oct. 31	08:53	22:22
Cape Town, South Africa			
	Oct. 01	04:24	16:49
	Oct. 31	03:47	17:13
London, UK			
	Oct. 01	06:01	17:39
	Oct. 31	06:52	16:36
Los Angeles, USA			
	Oct. 01	13:48	01:39
	Oct. 31	14:12	01:02
Nairobi, Kenya			
	Oct. 01	03:19	15:26
	Oct. 31	03:12	15:21
Sydney, Australia			
	Oct. 01	19:32	07:57
	Oct. 31	18:55	08:22
Tokyo, Japan			
	Oct. 01	20:36	08:26
	Oct. 31	21:02	07:47
Washington, DC, USA			
	Oct. 01	11:04	22:51
	Oct. 31	11:34	22:09
Wellington, New Zealand			
	Oct. 01	17:55	06:26
	Oct. 31	17:09	07:00

NB: *the times given are in Universal Time (UT)*

The Moon's phases and ages

Northern hemisphere

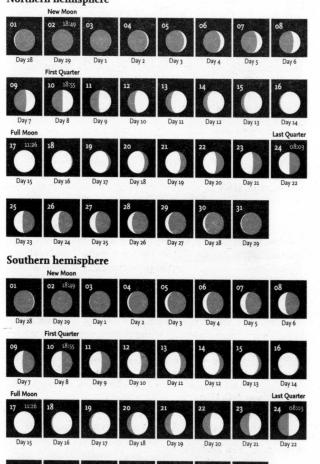

Southern hemisphere

The Moon

The Moon in October
On October 2 there is an annular solar eclipse, visible over the
Pacific Ocean and ending over the southernmost tip of South
America. The next day the Moon is 1.8° south of **Mercury** (which
is just mag. -1.6). On October 5 the Moon is 3.0° south of bright
Venus (mag. -3.9). By October 14 it is 0.1° north of **Saturn**. On
October 15 the Moon is 0.6°N of **Neptune** (which is just mag. 7.8)
and on October 19 it is 4.5° north of the slightly brighter **Uranus**
(at mag. 5.6). On October 23 the Moon is 3.9° north of **Mars**
(mag. 0.2).

Hunter's Moon
In the northern hemisphere of the Old World, October was
the month in which people prepared for the coming winter by
hunting wild animals, slaughtering livestock and preserving
meat for food. This caused the Full Moon in October to become
known as the 'Hunter's Moon'. Every three years, however,
the first Full Moon after the autumnal equinox fell, not in
September, but early in October, when it was also known as the
'Harvest Moon'. The October Full Moon was also known as the
'Dying Grass Moon' and the 'Blood' or 'Sanguine Moon'. ('Blood
Moon' is also the term sometimes applied to the Moon during a
lunar eclipse.)

   In the New World there was a great variety of names. Many
were related to the changes in autumn, such as 'Leaf-falling
Moon', 'Falling Leaves Moon' or simply 'Fall Moon'. To the
Algonquin of the Northeast and Great Lakes area it was the
'White Frost on Grass Moon'. The Assiniboine of the Northern
Plains had a rather different sort of name. To them it was the
'Joins Both Sides Moon'.

Édouard Albert Roche
The French astronomer and mathematician Édouard Albert
Roche (1820–1883) was born at Montpellier on 17 October 1820.
He is probably best known for his work in celestial mechanics,
and especially for the Roche Limit, named after him.

The Roche Limit

The Roche Limit (also sometimes known as the Roche Radius) is the distance from a celestial body, such as a planet, at which a body orbiting that body, for example a satellite or moon, that is held together only by its own gravity, will be disrupted by the tidal forces exerted by the primary body. Inside the Roche Limit the individual particles will tend to form a ring, whereas outside that limit they will tend to coalesce into a larger body. Bodies (natural or artificial) that are held together by mechanical forces, rather than just their own gravity may exist within the Roche Limit. Artificial satellites, for example, are able to orbit inside the Roche Limit, because their components are held together by purely mechanical forces.

The 160-km diameter lunar crater (or walled plain) Roche, located on the far side of the Moon, is named in his honour. The southern rim of this crater is interrupted by the 84-km impact crater Pauli.

Édouard Albert Roche.

The farside crater Roche and the smaller impact crater Pauli that lies on its southern rim.

O

Calendar for October

02	18:45	Annular solar eclipse (Pacific)
02	18:49	New Moon
02	19:39	Moon at apogee = 406,516 km (most distant of the year)
02–Nov.07		Orionid meteor shower
03	00:02	Mercury (mag. -1.6) 1.8°N of the Moon
05	20:26	Venus (mag. -3.9) 3.0°N of the Moon
06		Daylight Saving Time begins (parts of Australia)
06–10		Draconid meteor shower
08–09		Draconid meteor shower maximum
10–11		Southern Taurid meteor shower maximum
10	18:55	First Quarter
14	18:13	Saturn (mag. 0.7) 0.1°S of the Moon An occultation of Saturn is visible from Asia and eastern South Africa
15	17:32	Neptune (mag. 7.8) 0.6°S of the Moon
17	11:26	Full Moon
17	22:50	Moon at perigee = 357,175 km
19	15:52	Uranus (mag. 5.6) 4.5°S of the Moon
20–Dec.10		Northern Taurid meteor shower
21–22		Orionid meteor shower maximum
21	08:04	Jupiter (mag. -2.6) 5.8°S of the Moon
23	19:55	Mars (mag. 0.2) 3.9°S of the Moon
24	08:03	Last Quarter
29	22:50	Moon at apogee = 406,161 km

October 14–15 • *High in the northeast, the waxing gibbous Moon passes Saturn (as seen from Sydney).*

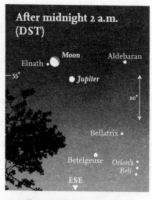

October 21 • *The Moon with Jupiter and Elnath, high in the east-southeast (as seen from central USA).*

October 23–24 • *The Moon is almost at Last Quarter when it passes between Pollux and Mars (as seen from London).*

October 24 • *The Moon is almost at Last Quarter, when it lines up with Castor, Pollux and Mars (as seen from Sydney).*

O

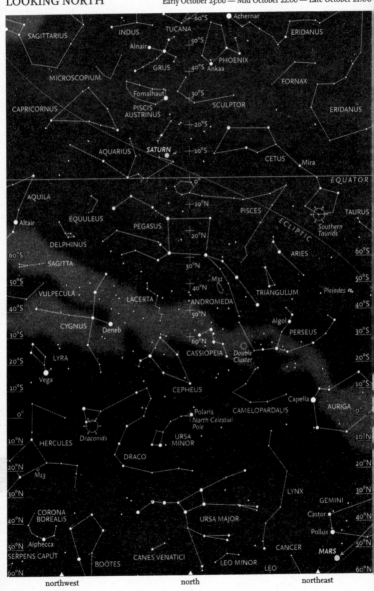

October – Looking North

The Milky Way arches across the northern sky, running from *Auriga* in the east to *Cygnus* and *Aquila* in the west. The constellations of *Perseus*, *Cassiopeia* and *Cepheus* lie along it. For observers in middle northern latitudes, these constellations are high overhead. (To observers in the far north, Cassiopeia is near the zenith.) *Andromeda* is high in the north, beyond Perseus and the other side of the Milky Way. The two small constellations of *Triangulum* and *Aries* are to the south of it. In the northwest, the constellation of *Lyra* lies farther south, clear of the Milky Way.

Between Cassiopeia and Cygnus lies the zig-zag of faint stars that form the small constellation of *Lacerta*, which is often difficult to recognize because it lies across the Milky Way. Two of the stars forming the Summer Triangle, *Deneb* (α Cygni) and *Vega* (α Lyrae) are readily visible, but the third, *Altair* (α Aquilae), is approaching the northwestern horizon. So too, much farther north, is the constellation of *Hercules*. The head of *Draco* is at about the same altitude as *Polaris* in *Ursa Minor*, as is *Capella* (α Aurigae). The large constellation of *Ursa Major* is now directly beneath the Pole, visible above the horizon to the north. *Castor* and *Pollux* (α and β Geminorum, respectively) are in the northeast, while the northernmost stars of *Boötes* and the circlet of *Corona Borealis* are in the northwest. The brightest star in Boötes, orange-tinted Arcturus, is below the horizon early in the night.

The long chain of faint stars that is the constellation of *Lynx* runs almost vertically between Ursa Major and *Gemini*, with the other faint constellation of *Camelopardalis* between it and Perseus.

O

This month, two of the northern circumpolar constellations, Cepheus and Cassiopeia, are almost at the meridian.

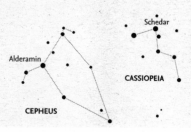

LOOKING SOUTH

Early October 23:00 — Mid October 22:00 — Late October 21:00

southeast   south   southwest

October – Looking South

The Great Square of *Pegasus* is now on the meridian due north, with the constellation of *Andromeda* stretching away from it in the northeast. The star at the northeastern corner of the Great Square is actually *Alpheratz* (α Andromedae). The two parts of the zodiacal constellation of *Pisces* are to the south and east of Pegasus. *Aquarius* and the non-zodiacal constellation of *Cetus* are slightly farther south while *Capricornus* is sinking in the west.

South of Aquarius is *Pisces Austrinus* with its single bright star, *Fomalhaut*, while straddling the meridian slightly east of it is the faint constellation of *Sculptor*. Below (south of) these two constellations are the roughly cross-shaped constellation of *Grus* with, on the other side of the meridian, *Phoenix*. Just below the constellation of Phoenix is *Achernar* (α Eridani), the bright star that ends the long winding constellation of *Eridanus* that actually begins near Rigel in Orion. Below Achernar is the triangular constellation of *Hydrus*, with the *Small Magellanic Cloud* (SMC) on its northwestern side. At about the same altitude towards the west is the constellation of *Pavo* with *Peacock* (α Pavonis).

The *Large Magellanic Cloud* (LMC) lies southeast of Hydrus, partly in each of the two faint constellations of *Dorado* and *Mensa*. The Milky Way lies in the east, where portions of *Sagittarius* and *Scorpius* are clearly visible. Roughly level with the 'Sting' of Scorpius are the constellations of *Ara* and *Triangulum Australe*. Farther east are the undistinguished constellations of *Apus*, *Chamaeleon* and *Volans*. Even farther to the east is brilliant *Canopus* (α Carinae), the second brightest star in the sky (after Sirius) at mag. -0.6.

For observers at about 30° south (roughly the latitude of Sydney in Australia), the two brightest stars of *Centaurus* (*Rigil Kentaurus* and *Hadar*) are brushing the horizon, while the small constellation of *Crux* is lost below it. Only observers even farther south will be able to see that constellation in full, together with the constellation of *Lupus*, the full extent of Centaurus and the large constellation of *Vela*, with *Puppis* in the southeast.

O

The Voyager probes
Following the success of the Pioneer spaceprobes, NASA
designed the Voyager program to take advantage of the
favourable positions of the planets Jupiter and Saturn. The two
spacecraft were launched in 1977, *Voyager 2* being actually the
first to be launched. *Voyager 1* was set on a faster track, primarily
so that it could make observations of Saturn's satellite Titan,
which would be favourably placed. It made gravity-assist passes
of Jupiter and Saturn. (These manoeuvres actually took it out
of the plane of the ecliptic, so it could not continue on to make
observations of any other objects.) Both spaceprobes made
successful observations of Jupiter and Saturn, with the data being
sent back to Earth. The trajectory for *Voyager 2* (launched first)
was designed so that the probe could be sent on to Uranus and
Neptune. It also obtained gravity assists from the two gas giants
to enable it to travel onwards to the outer planets.

Miranda, the satellite of Uranus as viewed by Voyager 2. *The
'racetrack-like' feature on the left is known as a corona. Also visible is
a feature now known as 'The Chevron' and, on the limb, the cliffs of
Verona Rupes.*

The immensely high icy scarp of Verona Rupes. These cliffs appear to be 20 km in height, making them the highest in the Solar System.

The spacecraft obtained many observations of Jupiter's highly complex meteorology. They observed Jupiter's main satellites (the Galilean moons) in detail and, in doing so, discovered the volcanic activity on Io, which has proved to be the most volcanically active object in the Solar System. At Saturn, apart from Voyager's first observations of Titan, they carried out extensive observations of the ring system, discovering the existence of numerous 'ringlets', 'braided' rings, and how the numerous satellites influence the nature of the rings. This included the discovery of 'shepherd' satellites that govern the structure of some of the individual rings.

Voyager 2 is the only spacecraft to have visited the two outermost gas giant (more properly, the 'ice giant') planets. It flew by Uranus in 1986 and Neptune in 1989. At Uranus it found a very extensive magnetic field, and no less than ten additional satellites. (An eleventh moon was then found in 1999 after additional analysis of *Voyager 2*'s data.) It provided dramatic images of Miranda, the innermost of the large satellites. Miranda has an extremely varied topography with some unique features. It also exhibits what is probably the highest cliff in the Solar System, known as Verona Rupes, which is believed to be 20 km high. Miranda, like the other satellites of Uranus, orbits the planet in its equatorial plane, which is highly inclined to the ecliptic. (The axial tilt is 97.77°.)

O

At Neptune, *Voyager 2* observed a 'Great Dark Spot' in the southern hemisphere, somewhat comparable to Jupiter's Great Red Spot. (Similar features, including major storms have since been observed by the Hubble Space Telescope in the northern hemisphere.) The spacecraft also revealed that Neptune has a system of narrow rings – thought, before the encounter, to be 'arcs' rather than true rings. (The nine rings forming the ring system around Uranus had been known since an occultation on 10 March 1977.) The planet proved to have a more complex and dynamic meteorology than Uranus, with fast-moving clouds. Some of the wind systems reach the highest speeds observed in the Solar System, achieving velocities of almost 2,200 kph.

The two *Voyager* spacecraft have subsequently continued their journeys outwards, and have both now passed the heliopause, the theoretical boundary where the solar wind is halted by the interstellar medium. This is at a distance of about 120 AU.

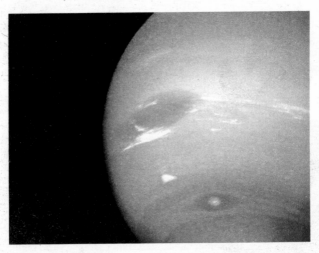

Neptune as viewed by Voyager 2, *showing the Great Dark Spot, bordered by high clouds; the fast-moving white patch of cloud, known as 'Scooter'; and the Small Dark Spot.*

November

November – Introduction

Three meteor streams begin in earlier months, but continue into November. The **Southern Taurid** shower, which began as long ago as September 10, may continue until November 20. Similarly, the **Orionids**, which began on October 2, continue until November 7. (Both of these showers are described in earlier months.) The third of these enduring meteor streams, the **Northern Taurid** shower, begins in late-October (October 20), and reaches its maximum – although with just a low rate of about five meteors per hour – on November 12–13. This is when the Moon is waxing gibbous (Day 11–12 of the lunation), so conditions are particularly poor. The shower gradually tails off, ending around December 10. There is an apparent 7-year periodicity in fireball activity, but 2024 is unlikely to be a peak year.

Far more striking, however, is the major shower of the **Leonids**, which have a relatively short period of activity (November 6–30), with maximum on November 17–18. This shower is associated with Comet 55P/Tempel-Tuttle, which has an orbital period of 33.22 years. The shower has shown extraordinary activity on various occasions with many thousands of meteors per hour. High rates were seen in 1999, 2001 and 2002 (reaching about 3000 meteors per hour) but have fallen dramatically since then. The rate in 2024 is likely to be about 15 per hour. Because of the radiant's location, this shower is best seen from the northern hemisphere. These meteors are the fastest shower meteors ever recorded (about 70 km per second) and often leave persistent trains. Apart from the sheer numbers occasionally seen, the shower is very rich in faint meteors. In 2024, maximum is when the Moon is waning gibbous (Days 16–17 of the lunation), so conditions are not very favourable.

There is an enigmatic, minor southern meteor shower that begins activity in late November (nominally November 28). This is the **Phoenicids**. Little is known of this shower, partly because the parent comet is believed to be the disintegrated comet D/1819 W1 (Blanpain). With no accurate knowledge of the location of the remnants of the comet, predicting the possible rate becomes little more than guesswork. The rate is variable and may rapidly increase if the orbit is nearby. There is a tendency for bright meteors to be frequent and all the meteors are fairly slow. The radiant is located within the constellation of Phoenix, not far from the border with Eridanus and the bright star *Achernar* (α Eridani).

The planets

Mercury reaches eastern elongation on November 16. *Venus* is very bright (mag.-4.0 to -4.2) and moving eastwards away from the Sun. *Mars* (mag. 0.1 to -0.4) moves eastwards in *Cancer*. *Jupiter* (mag.-2.7 to -2.8) is retrograding in *Taurus*. *Saturn* (mag. 0.8 to 0.9) is in *Aquarius*. *Uranus* (mag. 5.6) is in *Taurus* and comes to opposition on November 17 (see the chart below). *Neptune* (mag. 7.8) remains in *Pisces*.

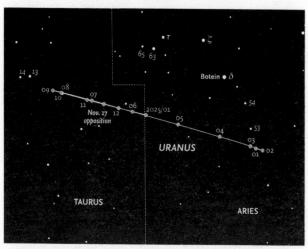

The path of Uranus in 2024. The planet comes to opposition on November 17 in Taurus. Stars down to magnitude 7.5 are shown.

Sunrise and sunset

City	Date	Sunrise	Sunset
Buenos Aires, Argentina			
	Nov. 01	08:52	22:23
	Nov. 30	08:34	22:51
Cape Town, South Africa			
	Nov. 01	03:46	17:14
	Nov. 30	03:28	17:41
London, UK			
	Nov. 01	06:54	16:34
	Nov. 30	07:43	15:56
Los Angeles, USA			
	Nov. 01	14:13	01:01
	Nov. 30	14:40	00:44
Nairobi, Kenya			
	Nov. 01	03:12	15:21
	Nov. 30	03:16	15:27
Sydney, Australia			
	Nov. 01	18:54	08:23
	Nov. 30	18:37	08:50
Tokyo, Japan			
	Nov. 01	21:03	07:46
	Nov. 30	21:31	07:28
Washington, DC, USA			
	Nov. 01	11:35	22:08
	Nov. 30	12:07	21:47
Wellington, New Zealand			
	Nov. 01	17:07	07:01
	Nov. 30	16:43	07:36

NB: the times given are in Universal Time (UT)

The Moon's phases and ages

Northern hemisphere

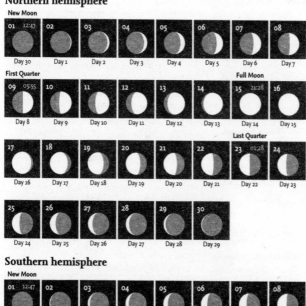

Southern hemisphere

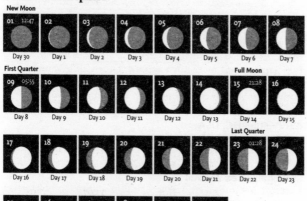

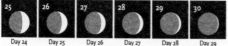

The Moon

The Moon in November

On November 3, the Moon is 2.1° south of **Mercury** (which is mag. -0.3). On November 4 the Moon is 0.1° south of **Antares** and the next day is 3.1°S of bright **Venus** (mag. -4.0). By November 12 the Moon is 0.6° north of faint **Neptune** (which is only mag. 7.8). On November 16 the Moon is 4.4° north of **Uranus** (which is at mag. 5.6). On November 17 the Moon is 10.3° north of **Aldebaran**, later that day it is 5.5° north of bright **Jupiter** (mag. -2.8). By November 20, the Moon is 1.9° south of **Pollux**, later that day it is 2.4° north of **Mars** in **Cancer**. By November 22, the Moon passes 2.7° north of **Regulus** between it and **Algieba**. On November 27 the Moon is 0.4° north of **Spica** in **Virgo**.

Beaver Moon

In North America, the Full Moon of November has come to be called 'Beaver Moon', because beavers become particularly active at this time, preparing their lodges and food supplies for winter, but this applies to a few only of the North-American tribes. Most see it as marking the beginning of heavy frosts, with names such as 'Freezing River Maker Moon', 'Freezing Moon', 'Rivers Begin to Freeze Moon', and 'Frost Moon', although to the Haida of Alaska, where snow lies for many months, it was the 'Snow Moon'.

In the Old World it was sometimes known as the 'Frosty Moon' or even, occasionally, as the 'Oak Moon', although this last term was more often applied to the Full Moon in December. If it was the last Full Moon before the winter solstice it was also sometimes known as the 'Mourning Moon'.

Edwin Hubble

The famous American astronomer Edwin Hubble (1889–1953) was born in Marshfield, Missouri on 20 November 1889. Following studies in Chicago and Oxford, and a period in the United States Army, Hubble became a staff member of the Carnegie Institution for Science's Mount Wilson Observatory in California. There, he had the use of the newly-comissioned 100-inch Hooker telescope, with which he established that many 'nebulae' were actually galaxies – isolated star systems – beyond the Milky Way. At the time, the Milky Way was considered to comprise the whole universe. Hubble was the first astronomer to use the new 200-inch telescope at Mount Palomar. Hubble went on to establish that there was a relationship between the distance to galaxies and the redshift that was found in their radial velocities, and thus between distance and velocity of recession. This relationship came to be known as the 'Hubble Law'. It has recently been pointed out that the Belgian astronomer, the Reverend Georges Lemaître had suggested such a law, and that this work preceeded that of Hubble. As a result, some authors describe the relationship as the 'Lemaître-Hubble Law'.

The 81 km diameter eroded and partly flooded lunar crater Hubble, located close to the east-northeastern limb of the Moon, is named in honour of Edwin Hubble as is the Hubble Space Telescope. A 91-km lunar crater on the farside of the Moon has been named after Georges Lemaître.

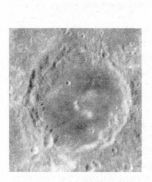

An Apollo 16 image of the 81-km diameter crater Hubble.

A famous portrait of the astronomer Edwin Hubble, taken in 1931.

N

Calendar for November

01	12:47	New Moon
03		Summer Time ends (UK reverts to GMT)
03	07:36	Mercury (mag. -0.3) 2.1°N of the Moon
04	01:06	Antares 0.1°N of the Moon
05	00:15	Venus (mag. -4.0) 3.1°N of the Moon
06–30		Leonid meteor shower
09	05:55	First Quarter
11	01:43	Saturn (mag. 0.9) 0.1°S of the Moon An occultation of Saturn is visible from Central America and northern South America
12–13		Northern Taurid meteor shower Maximum
12	02:25	Neptune (mag. 7.8) 0.6°S of the Moon
14	11:16	Moon at perigee = 360,109 km
15	21:28	Full Moon
16	01:13	Uranus (mag. 5.6) 4.4°S of the Moon
16	08:09	Mercury at greatest elongation (22.6°E, mag. -0.3)
17	02:02	Aldebaran 10.3°S of the Moon
17	02:45	Uranus at opposition (mag. 5.6)
17	14:53	Jupiter (mag. -2.8) 5.5°S of the Moon
17–18		Leonid meteor shower maximum
20	02:44	Pollux 1.9°N of the Moon
20	21:09	Mars (mag. -0.3) 2.4°S of the Moon
22	21:29	Regulus 2.7°S of the Moon
23	01:28	Last Quarter
26	11:56	Moon at apogee = 405,314 km
27	12:16	Spica 0.4°S of the Moon
28–Dec.09		Phoenicid meteor shower

November 3–5 • he Moon passes Mercury, Antares, Sabik, Venus and the Cat's Eyes (as seen from Sydney).

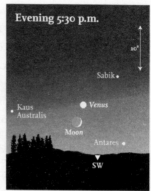

November 4 • The Moon is close to Venus and surrounded by Antares, Sabik and Kaus Australis (as seen from central USA).

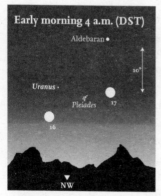

November 16–17 • The Full Moon passes Uranus (mag. 5.6), the Pleiades and Aldebaran (as seen from Sydney).

November 19–20 • The Moon passes Castor, Pollux and Mars (as seen from central USA).

N

November – Looking North

For observers at mid-northern latitudes, the Milky Way is now high in the north and northwest. **Cassiopeia** is not far from the zenith and **Cepheus** is 'on its side' in the northwest. **Capella** (α Aurigae) and the whole constellation of **Auriga** are high in the northeast. One of the stars of the Summer Triangle, Altair in Aquila, is now disappearing below the horizon. The other two: **Deneb** in **Cygnus** and **Vega** in **Lyra** are still visible, but low towards the northwestern horizon. **Eltanin** (γ Draconis) the brightest star in the 'head' of **Draco**, is at about the same altitude as Vega. **Ursa Minor**, together with **Polaris** itself, is slightly higher in the sky.

The end of the 'tail' of **Ursa Major** (that is, the star **Alkaid**, η Ursae Majoris), is almost due north, although the body of the constellation has begun to swing round into the northeast. Most of the fainter stars in the constellation to the south and east are easily visible, as is the undistinguished constellation of **Canes Venatici** to the southwest of it. Another insignificant constellation, often ignored, is **Leo Minor** to the southeast of Ursa Major. This consists of little more than three faint stars. **Gemini**, with the two bright stars, **Castor** and **Pollux**, is off to the east, with the line of stars forming **Lynx** between that constellation and Ursa Major. To the west of Ursa Major is **Hercules** although some of its stars are very low, close to the horizon, and likely to be difficult to see. Slightly towards the south are the northernmost stars of **Boötes**, although Arcturus itself is below the northern horizon.

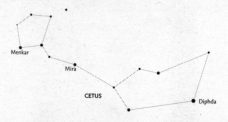

This time of the year, the constellation of Cetus and its famous variable star, Mira (o Ceti), is clearly visible. Because the celestial equator crosses the constellation, it is visible from almost every latitude.

southeast   south   southwest

218

November – Looking South

Several major constellations dominate the southern sky. There is *Pegasus* with the Great Square and *Andromeda* above and to the east of it. *Pisces* straddles the meridian, with the 'Circlet' to the south of the Great Square. The zodiacal constellation of *Aquarius* is sinking in the west, but is still clearly visible. To the east is *Taurus* and orange-tinted *Aldebaran.* Below (south) of Pisces the whole of *Cetus* and its famous variable star, *Mira* (o Ceti), is clearly visible. *Orion* is rising in the east, and the beginning of *Eridanus* at the stars λ Eridani and *Cursa* (β Eridani) near *Rigel* (β Orionis) is visible as are most of the stars in the long chain that forms this constellation as it winds its way south. The brightest star, *Achernar* (α Eridani), at the very end of the constellation, is on the horizon for observers at a latitude of 30° north. This constellation once ended at *Acamar,* (θ Eridani) the moderately bright star south of the small constellation of *Fornax,* and originally called Achernar until the constellation was extended and the name transferred to the current, brighter star.

Immediately north and west of Achernar is *Phoenix* and between that and Aquarius lie the two constellations of *Sculptor* and *Piscis Austrinus,* the latter with its solitary bright star, *Fomalhaut.* Below Piscis Austrinus is the constellation of *Grus.* Both the Magellanic Clouds are clearly seen. The Small Cloud *(SMC)* on the border between *Tucana* and *Hydrus* is almost on the meridian, and the Large Cloud *(LMC)* is farther east, bordered by *Dorado* and *Mensa. Canopus* (α Carinae) is at almost the same altitude, and the stars of *Puppis* are farther east. South of Canopus are the stars of *Vela* and *Carina. Crux* is now more-or-less 'upright', just east of the meridian. *Rigil Kentaurus* and *Hadar* are on the other side of the meridian. All these significant stars are right on the horizon for observers at 30° south, and only visible later in the night and later in the month. The zodiacal constellation of *Scorpius* is disappearing in the southwest and the stars of the sprawling constellation of *Centaurus* and those of *Lupus* are low in the sky.

N

Pioneer 10
On 2 March 1972, NASA launched the ***Pioneer 10*** spaceprobe. It was the very first spacecraft destined to study the objects in the outer Solar System, and became the first object to fly past Jupiter, at a closest distance of 132,252 km on 6 November 1973. One of its images is shown here.

An image of Jupiter obtained by Pioneer 10 *in November 1976. The Great Red Spot is just visible on the eastern limb, with a white oval farther west.*

Pioneer 10 passed the orbit of Saturn in 1976, that of Uranus in 1979 and Neptune's orbit on 13 June 1983.

    Pioneer 10 became the first spaceprobe to leave the Solar System. Its weak signal was used for training purposes after March 1997 and the mission officially ended that month. The last useful telemetry from the instruments on board came in April 2002. Communications with the probe were lost on 23 January 2003, when the signal was extremely weak and unable to provide any usable data. The signal was finally lost because of a loss of power to the radio transmitter.

December

December – Introduction

The last major meteor shower of the year is the *Geminid* shower, which is visible over the period December 4–20 and comes to maximum on December 14–15. In 2024 at maximum the Moon is just before and at Full Moon, so conditions are very unfavourable. It is one of the most active showers of the year, and in some years is the strongest, with a peak rate of around 100 meteors per hour. It is the one major shower that shows good activity before midnight. The meteors have a much higher density than most meteors (which are derived from cometary material). It was eventually established that the Geminids and the asteroid Phaethon had similar orbits. (Phaethon is the named asteroid that comes closest to the Sun at perihelion. There are unnamed numbered objects with closer perihelia.)

The Geminids are assumed to consist of denser, rocky material shed from Phaethon. They are slower than most other meteors and often seem to last longer. The brightest frequently break up into numerous luminous fragments that follow similar paths across the sky.

There is a second, minor, northern shower: the *Ursids*, which is active December 17–26, peaking on December 22–23. The maximum rate is generally approximately 10 meteors per hour, although much higher rates have been observed. The radiant is in Ursa Minor, near the star Kochab (β UMi). The stream is associated with Comet 8P/Tuttle, and is an extremely dense cluster of particles.

There are two southern meteor showers that may be seen in December. The *Phoenicid* shower begins in late November (November 28), and continues into December, reaching its weak maximum on December 2. This is immediately after New Moon, so observing conditions could not be more favourable. The other shower, the *Puppid Velids*, has its radiant on the border between the two constellations of Puppis and Vela. The shower begins on December 1, lasting until December 15, with maximum on December 7. (In 2024, this is just before First Quarter, so conditions are reasonably favourable.) It is a weak shower with a maximum rate of about 10 meteors per hour, although bright meteors are frequently visible.

The planets

Mercury reaches western elongation on December 25. *Venus* (mag. -4.2 to -4.5) moves rapidly across *Capricornus*. *Mars* begins retrograde motion on December 8. *Jupiter* (mag. -2.8) comes to opposition on December 7 (see the chart on page 229). *Saturn* (mag. 1.0 to 1.1) is in *Aquarius*. *Uranus* (mag. 5.6) is in *Taurus*, and *Neptune* (mag. 7.9) remains in *Pisces*, where it has been all year. Minor planet *(15) Eunomia* just reaches mag. 8.0 at opposition on December 14 (see the charts below).

A finder chart for the position of (15) Eunomia at its opposition on December 14. The white cross marks the position of Eunomia at the day of its opposition. The grey area is shown in more detail on the chart below, with stars down to magnitude 9.0.

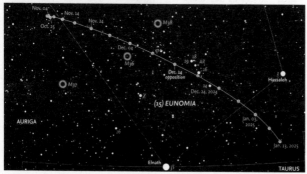

D

Sunrise and sunset

City	Date	Sunrise	Sunset
Buenos Aires, Argentina			
	Dec. 01	08:34	22:51
	Dec. 31	08:43	23:09
Cape Town, South Africa			
	Dec. 01	03:28	17:42
	Dec. 31	03:37	18:00
London, UK			
	Dec. 01	07:44	15:55
	Dec. 31	08:07	16:01
Los Angeles, USA			
	Dec. 01	14:41	00:44
	Dec. 31	14:59	00:53
Nairobi, Kenya			
	Dec. 01	03:16	15:27
	Dec. 31	03:30	15:41
Sydney, Australia			
	Dec. 01	18:37	08:51
	Dec. 31	18:47	09:09
Tokyo, Japan			
	Dec. 01	21:32	07:28
	Dec. 31	21:50	07:37
Washington, DC, USA			
	Dec. 01	12:08	21:47
	Dec. 31	12:27	21:56
Wellington, New Zealand			
	Dec. 01	16:42	07:37
	Dec. 31	16:51	07:57

NB: *the times given are in Universal Time (UT)*

The Moon's phases and ages

Northern hemisphere

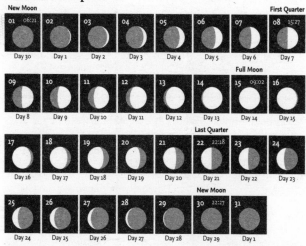

Southern hemisphere

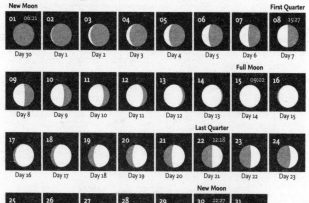

225

The Moon

The Moon in December
On December 1, the Moon is 0.1° south of *Antares* in *Scorpius*. The next day it is 5.0° south of *Mercury*. (Too close to the Sun to be visible.) By December 4 the Moon is 2.3° south of *Venus* (which is very bright at mag. -4.2) in *Sagittarius*. On December 8 it is 0.3° north of *Saturn* in *Aquarius*. The Moon passes 0.8° north of *Neptune* on December 9 and 4.4° north of *Uranus* on December 13. The next day, December 14, the Moon is 10.3° north of *Aldebaran* and later, 5.5° north of *Jupiter*. By December 17 the Moon is 2.1° south of *Pollux*. On December 18 the Moon is 0.9° north of *Mars*, which has begun to retrograde in *Cancer*. On December 20, the Moon passes 2.4° north of *Regulus*. On December 24 it is 0.2° north of *Spica* and on December 28, 0.1° south of *Antares*. One day before New, the Moon is 6.4° south of *Mercury*.

Cold Moon
In the northern hemisphere, the cold of winter begins to dominate life during December. Many of the names for the Full Moon in December on both sides of the Atlantic refer to the temperature, with terms such as 'Cold Moon', 'Winter Maker Moon' and 'Snow Moon' used in North America. The Cree of Canada had a rather strange name: 'Moon when the Young Fellow Spreads the Brush'.

The Japanese spacecraft *Hayabusa2* returned samples of the asteroid *Ryugu* to Earth on 6 December 2020. The probe has since been redirected to asteroid 1998 KY$_{26}$.

Among the Zuñi of New Mexico it was the 'Moon when the Sun has Travelled Home to Rest'. In Europe it was sometimes called the 'Moon before Yule' or the 'Wolf Moon', although that term was more commonly applied to the Full Moon in January.

Wilhelm Tempel

The German astronomer Ernst Wilhelm Leberecht Tempel, always known as Wilhelm Tempel (1821–1889) was born at Niedercunnersdorf, Saxony, on 4 December 1821. He initially worked at the Marseille Observatory but then moved to Italy. He discovered 21 comets, including the famous Comet 55P/Tempel-Tuttle, now known to be the source of the Leonid meteor shower (see page 208) and also Comet Tempel 1 (the periodic comet 9P/Tempel) that was the body studied by the *Deep Impact* spaceprobe (see page 124). The Main-Belt asteroid 3808 Tempel and a lunar crater are named after him.

Ernst Wilhelm Leberecht Tempel.

Comet Tempel 1
Wilhelm Tempel discovered this comet on 3 April 1867. The comet regularly passes close to Jupiter, which causes changes in the orbit and perihelion. For this reason the comet was 'lost' until 1967, when the astronomer Brian Marsden recalculated the orbit, taking such close approaches into account.

An image of Comet Tempel 1 in X-rays, as obtained by the Chandra *spacecraft.*

A Lunar Orbiter 4 image of the 43-km badly eroded lunar impact crater Tempel, named after the astronomer.

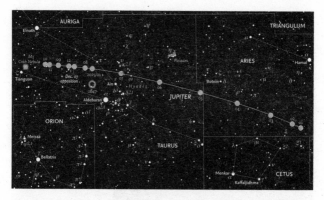

The path of Jupiter in 2024. Jupiter comes to opposition on December 7 in Taurus. Stars down to magnitude 6.5 are shown.

The Chinese lunar probe **Chang'e-5** landed in Oceanus Procellarum on 1 December 2020. It returned lunar samples to Earth on 16 December 2020, showing that lunar volcanism had persisted for much later than previously believed.

Calendar for December

01–15		Puppid Velid meteor shower
01	06:21	New Moon
01	07:26	Antares 0.1°N of the Moon
02		Phoenicid meteor shower maximum
02	02:09	Mercury (mag. 2.7) 5.0°N of the Moon
04–20		Geminid meteor shower
04	22:40	Venus (mag. -4.2) 2.3°N of the Moon
07	20:58	Jupiter at opposition (mag. -2.8)
08	09:19	Saturn (mag. 1.0) 0.3°S of the Moon
08	15:27	First Quarter
09	09:19	Neptune (mag. 7.9) 0.8°S of the Moon
12	13:20	Moon at perigee = 365,361 km
13	09:34	Uranus (mag. 5.6) 4.4°S of the Moon
14–15		Geminid meteor shower maximum
14	05:31	Minor planet (15) Eunomia at opposition (mag. 8.0)
14	12:30	Aldebaran 10.3°S of the Moon
14	19:32	Jupiter (mag. -2.8) 5.5°S of the Moon
15	09:02	Full Moon
17–26		Ursid meteor shower
17	12:49	Pollux 2.1°N of the Moon
18	08:49	Mars (mag. -0.9) 0.9°S of the Moon
20	06:18	Regulus 2.4°S of the Moon
21	09:21	December equinox
22	22:18	Last Quarter
22–23		Ursid meteor shower maximum
24	07:25	Moon at apogee = 404,485 km
24	20:11	Spica 0.2°S of the Moon
25	02:30	Mercury at greatest elongation (22.0°W, mag. -0.4)
28	15:17	Antares 0.1°N of the Moon
29	04:21	Mercury (mag. -0.4) 6.4°N of the Moon
30	22:27	New Moon

December 8 • *High in the northwest, the Moon is close to Saturn (as seen from Sydney).*

December 13–15 • *The Moon is nearly Full, when it passes the Pleiades, Aldebaran, Jupiter and Elnath (as seen from London).*

December 17 • *Mars, the Moon, Pollux and Castor form a nice curved line in the eastern sky (as seen from London).*

December 19 • *The Moon is between Regulus and Algieba, due east (as seen from central USA).*

D

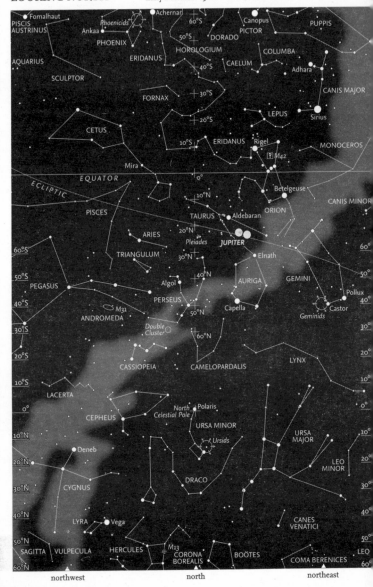

LOOKING NORTH

Early December 23:00 — Mid December 22:00 — Late December 21:00

PISCIS AUSTRINUS · Fomalhaut · Achernar · 60°S · Canopus · PUPPIS
Ankaa · *Phoenicids* · DORADO · PICTOR
PHOENIX · 50°S
AQUARIUS · ERIDANUS · HOROLOGIUM · 40°S · CAELUM · COLUMBA
SCULPTOR · FORNAX · 30°S · Adhara
CETUS · 20°S · LEPUS · CANIS MAJOR
· 10°S · ERIDANUS · Rigel · Sirius
Mira · M42 · MONOCEROS
ECLIPTIC · EQUATOR · 0° · Betelgeuse
· 10°N · CANIS MINOR
PISCES · TAURUS · Aldebaran · ORION
· 20°N · *Pleiades* · **JUPITER** · Elnath
ARIES · 30°N · AURIGA · GEMINI
TRIANGULUM · Algol · 40°N · Pollux
PEGASUS · PERSEUS · Capella · Castor
· 50°N · *Geminids*
ANDROMEDA · M31 · *Double Cluster* · 60°N
· CASSIOPEIA · CAMELOPARDALIS · LYNX
LACERTA · CEPHEUS · *North Celestial Pole* · Polaris · URSA MINOR
· *Ursids* · URSA MAJOR
· Deneb · LEO MINOR
CYGNUS · DRACO
LYRA · Vega · CANES VENATICI
SAGITTA · VULPECULA · HERCULES · M13 · CORONA BOREALIS · BOÖTES · COMA BERENICES · LEO

northwest · north · northeast

232

December – Looking North

For observers close to the equator, **Orion** is high overhead, to the east of the meridian, with **Taurus**, **Auriga** and **Gemini** way above the northern horizon. For observers at mid-northern latitudes, it is **Perseus** that is at the zenith, with **Andromeda** stretching off to the west and the Great Square of **Pegasus** in the northwest. Auriga with **Capella** is slightly to the east. Even farther east are the two bright stars of **Gemini**, **Castor** and **Pollux**.

The Milky Way runs from high in the northeast down to the northwest. Below Perseus is the distinctive shape of **Cassiopeia**, which lies within the Milky Way, and farther down is the zig-zag constellation of **Lacerta**, which is like **Cepheus**, in that both of them are partly within the band of stars. Even farther towards the northwest is **Cygnus** and the beginning of the Great Dark Rift near **Deneb**. The constellation of **Lyra**, with brilliant **Vega**, lies towards the meridian, away from the star clouds of the Milky Way. Much of the constellation of **Hercules** is clear of the northern horizon, together with some of the northernmost stars in **Boötes**.

Ursa Major is climbing in the northeast, and all the far-flung outlying stars are clearly visible. Below the 'tail' is the inconspicuous constellation of **Canes Venatici**, and some observers at high latitudes may even be able to glimpse some of the stars of **Coma Berenices**, low on the horizon in the northeast. Between Ursa Major and the Milky Way, the whole of the constellations of **Draco** and **Ursa Minor** are clearly visible as is Cepheus beyond them. The chain of stars forming the faint constellation of **Lynx** lies between Ursa Major and the bright stars of Gemini.

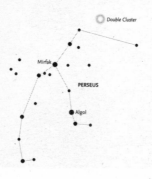

Double Cluster

Mirfak

PERSEUS

Algol

This time of the year, the constellation of Perseus, with the famous variable star Algol and the Double Cluster, is at the meridian.

D

233

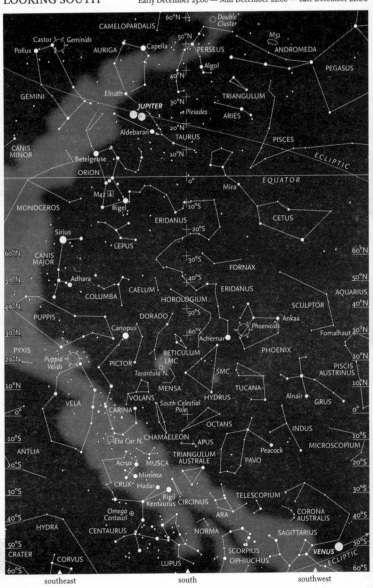

CAMELOPARDALIS
60°N
Double Cluster
Castor
Geminids
Pollux
Capella
AURIGA
PERSEUS
M31
ANDROMEDA
50°N
Algol
40°N
PEGASUS
GEMINI
Elnath
JUPITER
30°N
TRIANGULUM
Pleiades
ARIES
PISCES
20°N
Aldebaran
TAURUS
ECLIPTIC
CANIS MINOR
10°N
Betelgeuse
ORION
0°
EQUATOR
Mira
MONOCEROS
10°S
M42
Rigel
ERIDANUS
CETUS
Sirius
20°S
LEPUS
60°N
CANIS MAJOR
30°S
FORNAX
50°N
Adhara
40°S
ERIDANUS
AQUARIUS
40°N
COLUMBA
CAELUM
PUPPIS
HOROLOGIUM
SCULPTOR
Ankaa
30°N
Canopus
50°S
DORADO
Phoenicids
Fomalhaut
PYXIS
PICTOR
60°S
Achernar
20°N
Puppid Velids
RETICULUM
LMC
PHOENIX
PISCIS AUSTRINUS
10°N
Tarantula N.
SMC
0°
VELA
VOLANS
MENSA
TUCANA
Alnair
GRUS
CARINA
South Celestial Pole
HYDRUS
10°S
ANTLIA
Eta Car N.
OCTANS
INDUS
MICROSCOPIUM
20°S
Acrux
MUSCA
APUS
CHAMAELEON
PAVO
Peacock
30°S
Mimosa
CRUX
Hadar
TRIANGULUM AUSTRALE
CIRCINUS
TELESCOPIUM
40°S
Omega Centauri
Rigil Kentaurus
ARA
CORONA AUSTRALIS
HYDRA
CENTAURUS
NORMA
SAGITTARIUS
50°S
CRATER
LUPUS
SCORPIUS
VENUS
60°S
CORVUS
OPHIUCHUS
ECLIPTIC

southeast   south   southwest

234

December – Looking South

Orion has now risen well clear of the horizon and is in the northeast. Above it are *Taurus* and *Auriga*, with *Gemini* farther to the east. Within the Milky Way to the east of Orion, is the rather undistinguished constellation of *Monoceros*, which is without any distinct, bright stars. Below Orion is the small constellation of *Lepus*, and to its east, *Canis Major*, with brilliant *Sirius*, the brightest star in the sky (mag. -1.4). The long, winding constellation of *Eridanus* begins near *Rigel* in Orion and runs south, partly enclosing *Fornax* (which was once part of the larger constellation), until it ends at *Achernar* (α Eridani). Between Eridanus and Canis Major is the tiny, faint constellation of *Caelum* and the larger and brighter *Columba*.

South of Canis Major is the constellation of *Puppis*, once (with *Vela* and *Carina*) part of the large obsolete constellation of Argo Navis. *Canopus* (α Carinae) and Achernar are close to the horizon for observers at 30°N. West of Achernar is *Phoenix* and even farther west, below *Aquarius*, the constellation of *Piscis Austrinus* is slowly descending in the west. Between Piscis Austrinus in the west and Carina in the east lie several small constellations, the most conspicuous of which is *Grus*. There are also *Tucana* and *Hydrus*, with the *Small Magellanic Cloud* (SMC), then the *Large Magellanic Cloud* (LMC) in *Mensa* and *Dorado*, with the tiny constellation of *Reticulum* north of them. North of that again is the faint constellation of *Horologium*. Between Mensa and the Milky Way in Carina is the tiny constellation of *Volans*.

More faint constellations lie around the South Celestial Pole in *Octans*, notably *Chamaeleon* and *Apus*. To the west lies the larger *Pavo* and the rather undistinguished constellation of *Indus*. *Crux* has now swung round into the southeast. Both it and the small neighbouring constellation of *Musca* lie in the Milky Way. *Rigil Kentaurus* and *Hadar*, the two brightest stars in *Centaurus* are slightly farther south, with *Circinus* and *Triangulum Australe* between them and Pavo. Those in the far south are able to see all the stars in Centaurus and *Lupus*, together with the southernmost stars of *Scorpius* (the 'sting') and *Sagittarius*, as well as other constellations, such as *Norma*, *Ara*, *Telescopium* and *Corona Australis*.

D

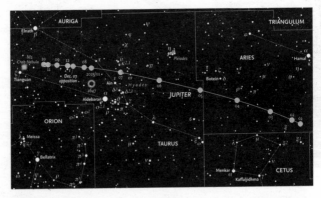

The path of Jupiter in 2024. Jupiter comes to opposition on December 7 in Taurus. Stars down to magnitude 6.5 are shown. The numbers relating to the positions of Jupiter in the diagram above correspond to the months of the year.

The Chinese lunar probe ***Chang'e-5*** landed in Oceanus Procellarum on 1 December 2020. It returned lunar samples to Earth on 16 December 2020, showing that lunar volcanism had persisted for much later than previously believed.

Dark-Sky Sites

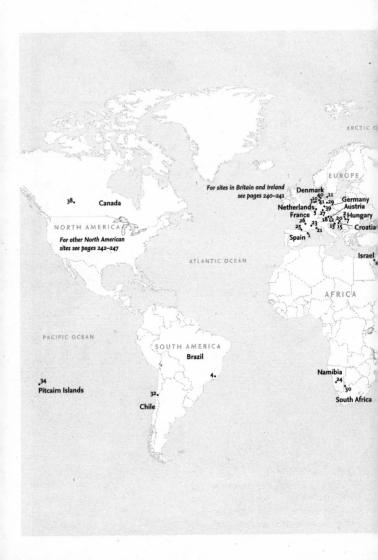

ARCTIC O

EUROPE

For sites in Britain and Ireland
see pages 240–241

Denmark

Germany

Austria

Netherlands

France

Hungary

Croatia

Spain

Israel

Canada

NORTH AMERICA

For other North American
sites see pages 242–247

ATLANTIC OCEAN

AFRICA

PACIFIC OCEAN

SOUTH AMERICA

Brazil

Namibia

Pitcairn Islands

Chile

South Africa

Dark-Sky Sites, Britain & Ireland

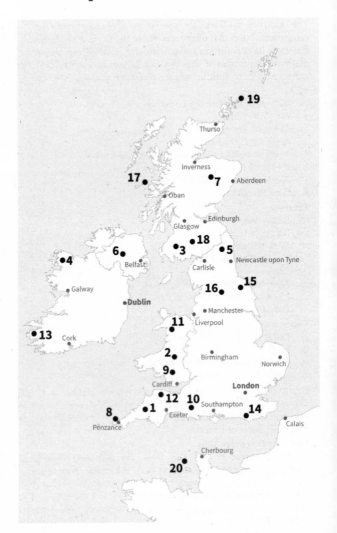

The International Dark-Sky Association (IDA) recognizes various categories of sites that offer areas where the sky is dark at night, free from light pollution and particularly suitable for astronomical observing. A number of sites in Great Britain and Ireland have been given specific recognition and are shown on the map. These are:

Parks

1   *Bodmin Moor Dark Sky Landscape*
2   *Elan Valley Estate*
3   *Galloway Forest Park*
4   *Mayo Dark Sky Park*
5   *Northumberland National Park and Kielder Water & Forest Park*
6   *OM Dark Sky Park & Observatory*
7   *Tomintoul and Glenlivet – Cairngorms*
8   *West Penwith*

Reserves

9   *Brecon Beacons National Park*
10   *Cranborne Chase*
11   *Eryri National Park (Snowdonia)*
12   *Exmoor National Park*
13   *Kerry*
14   *Moore's Reserve South Downs National Park*
15   *North York Moors National Park*
16   *Yorkshire Dales National Park*

Communities

17   *Coll* (Inner Hebrides, Scotland)
18   *Moffat*
19   *North Ronaldsay Dark Sky Island* (Orkney, Scotland)
20   *Sark* (Channel Islands)

Details of these sites and web links may be found at the IDA website: https://www.darksky.org/. Many of these sites have major observatories or other facilities available for public observing (often at specific dates or times).

Dark Sky Discovery Sites

In Britain there is also the *Dark Sky Discovery* organization. This gives recognition to smaller sites, again free from immediate light pollution, that are open to observing at any time. Some sites are used for specific, public observing sessions. A full listing of sites is at https://www.darkskydiscovery.org.uk/ but specific events are publicized locally.

US Dark-Sky Sites

International Dark Sky Association Sites

The International Dark-Sky Association (IDA) recognizes various categories of sites that offer areas where the sky is dark at night, free from light pollution and particularly suitable for astronomical observing. There are numerous sites in the United States, shown on the map and listed here.

Details of IDA are at: https://www.darksky.org/.

Information on the various categories and individual sites are at: https://www.darksky.org/our-work/conservation/idsp/

Many of these sites have major observatories or other facilities available for public observing (often at specific dates or times).

Parks

1   *AMC Maine Woods* (ME)
2   *Antelope Island State Park* (UT)
3   *Anza-Borrego Desert State Park* (CA)
4   *Arches National Park* (UT)
5   *Big Bend National Park* (TX)
6   *Big Bend Ranch State Park* (TX)
7   *Big Cypress National Preserve* (FL)
8   *Black Canyon of the Gunnison National Park* (CO)
9   *Bryce Canyon National Park* (UT)
10 *Buffalo National River* (AR)
11 *Canyonlands National Park* (UT)
12 *Cape Lookout National Seashore* (NC)
13 *Capitol Reef National Park* (UT)
14 *Capulin Volcano National Monument* (NM)
15 *Cedar Breaks National Monument* (UT)
16 *Chaco Culture National Historical Park* (NM)
17 *Cherry Springs State Park* (PA)
18 *Chiricahua National Monument* (AZ)
19 *Clayton Lake State Park* (NM)
20 *Copper Breaks State Park* (TX)
21 *Craters Of The Moon National Monument* (ID)
22 *Curecanti National Recreation Area* (CO)
23 *Dead Horse Point State Park* (UT)
24 *Death Valley National Park* (CA)
25 *Dinosaur National Monument* (CO)
26 *Dr. T.K. Lawless County Park* (MI)
27 *East Canyon State Park* (UT)
28 *El Morro National Monument* (NM)
29 *Enchanted Rock State Natural Area* (TX)
30 *Flagstaff Area National Monuments* (AZ)
31 *Florissant Fossil Beds National Monument* (CO)

Reserves

Sanctuaries

RASC Recognized Dark-Sky Sites

Canadian Dark-Sky Sites
The Royal Astronomical Society of Canada (RASC) has developed formal guidelines and requirements for three types of light-restricted protected areas: Dark-Sky Preserves, Urban Star Parks and Nocturnal Preserves. The focus of the Canadian Program is primarily to protect the nocturnal environment; therefore, the outdoor lighting requirements are the most stringent, but also the most effective. Canadian Parks and other areas that meet these guidelines and successfully apply for one of these designations are officially recognized. Many parks across Canada have been designated in recent years – see the list below and the RASC website: https://www.rasc.ca/dark-sky-site-designations.

Dark-Sky Preserves

1   *Au Diable Vert* (QC)

2   *Beaver Hills* (AB)

3   *Bluewater Outdoor Education Centre* (ON)

4   *Bruce Peninsula National Park* (ON)

5   *Cypress Hills* (SK/AB)

6   *Fundy National Park* (NB)

7   *Grasslands National Park* (SK)

8   *Jasper National Park* (AB)

9   *Kejimkujik National Park and National Historic Site* (NS)

10 *Killarney Provincial Park* (ON)

11 *Kouchibouguac National Park* (NB)

12 *Lakeland Provincial Park and Recreation Area* (AB)

13 *Lake Superior Provincial Park* (ON)

14 *Manitoulin Eco Park* (ON)

15 *McDonald Park* (BC)

16 *Mont-Mégantic International Dark-Sky Reserve* (QC)

17 *Mount Carleton Provincial Park* (NB)

18 *North Frontenac* (ON)

19 *Point Pelee National Park* (ON)

20 *Terra Nova National Park* (NL)

21 *Torrance Barrens Conservation Area* (ON)

22 *Wood Buffalo National Park* (AB)

Urban Star Parks

23 *Cattle Point* (BC)

24 *Irving Nature Park* (NB)

Nocturnal Preserves

25 *Ann and Sandy Cross Conservation Area* (AB)

26 *Old Man on His Back Prairie and Heritage Conservation Area* (SK)

Dark Sky Sites Around the World

Parks

1 *Albanyà* (Spain)

2 *Bükk National Park* (Hungary)

3 *De Boschplaat* (Netherlands)

4 *Desengano State Park* (Brazil)

5 *Eifel National Park* (Germany)

6 *Hehuan Mountain* (Taiwan)

7 *Hortobágy National Park* (Hungary)

8 *Iriomote-Ishigaki National Park* (Japan)

9 *Kozushima Dark Sky Island* (Japan) – also a Dark Sky Community

10 *Lauwersmeer National Park* (Netherlands)

11 *Møn and Nyord* (Denmark)

12 *Naturpark Attersee-Traunsee* (Austria)

13 *Petrova Gora-Biljeg* (Croatia)

14 *Ramon Crater Nature Reserve* (Israel)

15 *Vrani kamen* (Croatia)

16 *Wai-Iti* (New Zealand)

17 *Warrumbungle Dark Sky Park* (Australia)

18 *Winklmoosalm* (Germany)

19 *Yeongyang Firefly Eco Park* (South Korea)

20 *Zselic National Landscape Protection Area* (Hungary)

Reserves

21 *Alpes Azur Mercantour* (France)
22 *Aoraki Mackenzie* (New Zealand)
23 *Cévennes National Park* (France)
24 *NambiRand Nature Reserve* (Namibia)
25 *Pic du Midi* (France)
26 *Regional Natural Park of Millevaches in Limousin* (France)
27 *Rhön* (Germany)
28 *River Murray* (Australia)
29 *Westhavelland* (Germany)

Sanctuaries

30 *!Ae!Hai Kalahari Heritage Park* (South Africa)
31 *Aotea / Great Barrier Island* (New Zealand)
32 *Gabriela Mistral* (Chile)
33 *Niue* (New Zealand) – also a Dark Sky Community
34 *Pitcairn Islands* (UK)
35 *Stewart Island / Rakiura* (New Zealand)
36 *The Jump-Up* (Australia)

Communities

37 *Bisei Town, Ibara City* (Japan)
38 *Bon Accord* (Canada)
39 *Fulda, Hesse* (Germany)
40 *Pellworm Star Island* (Germany)
41 *Spiekeroog Star Island* (Germany)

Twilight Diagrams

Sunrise, sunset, twilight

For each individual month, we give details of sunrise and sunset times for nine cities across the world. But observing the stars is also affected by twilight, and this varies considerably from place to place. During the summer, especially at high latitudes, twilight may persist throughout the night and make it difficult to see the faintest stars. Beyond the Arctic and Antarctic Circles, of course, the Sun does not set for 24 hours at least once during the summer (and rise for 24 hours at least once during the winter). Even when the Sun does dip below the horizon at high latitudes, bright twilight persists throughout the night, so observing the stars is impossible.

There are three recognized stages of twilight: civil twilight, when the Sun is less than 6° below the horizon; nautical twilight, when the Sun is between 6° and 12° below the horizon; and astronomical twilight, when the Sun is between 12° and 18° below the horizon. Full darkness occurs only when the Sun is more than 18° below the horizon. During nautical twilight, only the very brightest stars are visible. (These are the stars that were used for navigation, hence the name for this stage.) During astronomical twilight, the faintest stars visible to the naked eye may be seen directly overhead, but are lost at lower altitudes. They become visible only once it is fully dark. The diagrams show the duration of twilight at the various cities. Of the locations shown, during the summer months there is full darkness at most of the cities, but it never occurs during the summer at the latitude of London. Observing conditions are most favourable at somewhere like Nairobi, which is very close to the equator, so there is not only little twilight, and a long period of full darkness, but there are also only slight variations in timing and duration throughout the year.

The diagrams also show the times of New and Full Moon (black and white symbols, respectively). As may be seen, at most locations during the year roughly half of New and Full Moon phases may come during daylight. For this reason, the exact phase may be invisible at one location, but be clearly seen elsewhere. The exact times of the events are given in the diagrams for each individual month.

Buenos Aires, Argentina – Latitude 34.7°S – Longitude 58.5°W

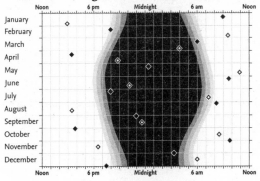

Cape Town, South Africa – Latitude 33.9°S – Longitude 18.5°E

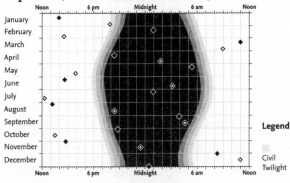

London, UK – Latitude 51.5°N – Longitude 0.2°W

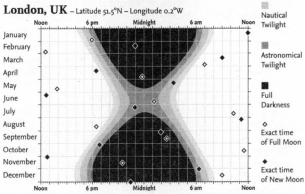

Legend

Civil Twilight

Nautical Twilight

Astronomical Twilight

Full Darkness

◇ Exact time of Full Moon

◆ Exact time of New Moon

Los Angeles, USA – Latitude 34.0°N – Longitude 118.2° W

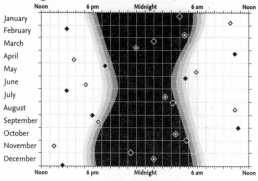

Nairobi, Kenya – Latitude 1.3°S – Longitude 36.8°E

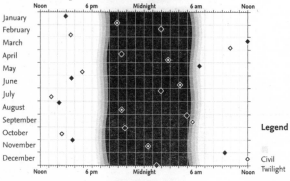

Sydney, Australia – Latitude 33.5°S – Longitude 151.2°E

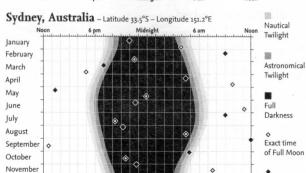

Legend

Civil Twilight

Nautical Twilight

Astronomical Twilight

Full Darkness

◇ Exact time of Full Moon

◆ Exact time of New Moon

Tokyo, Japan – Latitude 35.7°N – Longitude 139.8°E

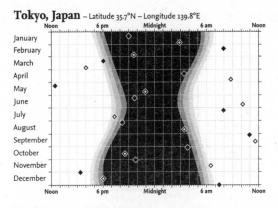

Washington, DC, USA – Latitude 38.9°N – Longitude 77.0°W

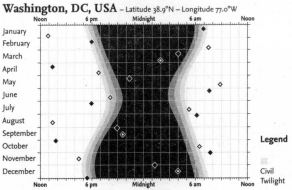

Wellington, New Zealand – Latitude 41.3°S – Longitude 174.8°E

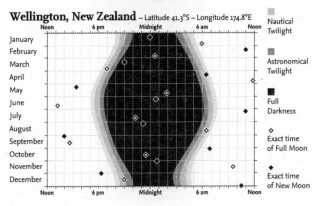

Legend

Civil Twilight

Nautical Twilight

Astronomical Twilight

Full Darkness

◇ Exact time of Full Moon

◆ Exact time of New Moon

Glossary and Tables

aphelion The point on an orbit that is farthest from the Sun.

apogee The point on its orbit at which the Moon is farthest from the Earth.

appulse The apparently close approach of two celestial objects; two planets, or a planet and star.

astronomical unit (AU) The mean distance of the Earth from the Sun, 149,597,870 km.

celestial equator The great circle on the celestial sphere that is in the same plane as the Earth's equator.

celestial sphere The apparent sphere surrounding the Earth on which all celestial bodies (stars, planets, etc.) seem to be located.

conjunction The point in time when two celestial objects have the same celestial longitude.
In the case of the Sun and a planet, superior conjunction occurs when the planet lies on the far side of the Sun (as seen from Earth). For Mercury and Venus, inferior conjunction occurs when they pass between the Sun and the Earth.

direct motion Motion from west to east on the sky.

ecliptic The apparent path of the Sun across the sky throughout the year.
Also: the plane of the Earth's orbit in space.

elongation The point at which an inferior planet has the greatest angular distance from the Sun, as seen from Earth.

equinox The two points during the year when night and day have equal duration.
Also: the points on the sky at which the ecliptic intersects the celestial equator.
The vernal (northern spring) equinox is of particular importance in astronomy.

gibbous The stage in the sequence of phases at which the illumination of a body lies between half and full.
In the case of the Moon, the term is applied to phases between First Quarter and Full, and between Full and Last Quarter.

inferior planet   Either of the planets Mercury or Venus, which have orbits inside that of the Earth.

magnitude   The brightness of a star, planet or other celestial body. It is a logarithmic scale, where larger numbers indicate fainter brightness. A difference of 5 in magnitude indicates a difference of 100 in actual brightness, thus a first-magnitude star is 100 times as bright as one of sixth magnitude.

meridian   The great circle passing through the North and South Poles of a body and the observer's position; or the corresponding great circle on the celestial sphere that passes through the North and South Celestial Poles and also through the observer's zenith.

nadir   The point on the celestial sphere directly beneath the observer's feet, opposite the zenith.

occultation   The disappearance of one celestial body behind another, such as when stars or planets are hidden behind the Moon.

opposition   The point on a superior planet's orbit at which it is directly opposite the Sun in the sky.

perigee   The point on its orbit at which the Moon is closest to the Earth.

perihelion   The point on an orbit that is closest to the Sun.

retrograde motion   Motion from east to west on the sky.

superior planet   A planet that has an orbit outside that of the Earth.

vernal equinox   The point at which the Sun, in its apparent motion along the ecliptic, crosses the celestial equator from south to north. Also known as the First Point of Aries.

zenith   The point directly above the observer's head.

zodiac   A band, stretching 8° on either side of the ecliptic, within which the Moon and planets appear to move. It consists of 12 equal areas, originally named after the constellation that once lay within it.

The Constellations

There are 88 constellations covering the whole of the celestial sphere. The names themselves are expressed in Latin, and the names of stars are frequently given by Greek letters (see page 258) followed by the genitive of the constellation name or its three-letter abbreviation. The genitives, the official abbreviations and the English names of the various constellations are included.

Name	Genitive	Abbr.	English name
Andromeda	Andromedae	And	Andromeda
Antlia	Antliae	Ant	Air Pump
Apus	Apodis	Aps	Bird of Paradise
Aquarius	Aquarii	Aqr	Water Bearer
Aquila	Aquilae	Aql	Eagle
Ara	Arae	Ara	Altar
Aries	Arietis	Ari	Ram
Auriga	Aurigae	Aur	Charioteer
Boötes	Boötis	Boo	Herdsman
Caelum	Caeli	Cae	Burin
Camelopardalis	Camelopardalis	Cam	Giraffe
Cancer	Cancri	Cnc	Crab
Canes Venatici	Canum Venaticorum	CVn	Hunting Dogs
Canis Major	Canis Majoris	CMa	Big Dog
Canis Minor	Canis Minoris	CMi	Little Dog
Capricornus	Capricorni	Cap	Sea Goat
Carina	Carinae	Car	Keel
Cassiopeia	Cassiopeiae	Cas	Cassiopeia
Centaurus	Centauri	Cen	Centaur
Cepheus	Cephei	Cep	Cepheus
Cetus	Ceti	Cet	Whale
Chamaeleon	Chamaeleontis	Cha	Chameleon
Circinus	Circini	Cir	Compasses
Columba	Columbae	Col	Dove
Coma Berenices	Comae Berenices	Com	Berenice's Hair
Corona Australis	Coronae Australis	CrA	Southern Crown
Corona Borealis	Coronae Borealis	CrB	Northern Crown

Name	Genitive	Abbr.	English name
Corvus	Corvi	Crv	Crow
Crater	Crateris	Crt	Cup
Crux	Crucis	Cru	Southern Cross
Cygnus	Cygni	Cyg	Swan
Delphinus	Delphini	Del	Dolphin
Dorado	Doradus	Dor	Dorado
Draco	Draconis	Dra	Dragon
Equuleus	Equulei	Equ	Little Horse
Eridanus	Eridani	Eri	River Eridanus
Fornax	Fornacis	For	Furnace
Gemini	Geminorum	Gem	Twins
Grus	Gruis	Gru	Crane
Hercules	Herculis	Her	Hercules
Horologium	Horologii	Hor	Clock
Hydra	Hydrae	Hya	Water Snake
Hydrus	Hydri	Hyi	Lesser Water Snake
Indus	Indi	Ind	Indian
Lacerta	Lacertae	Lac	Lizard
Leo	Leonis	Leo	Lion
Leo Minor	Leonis Minoris	LMi	Little Lion
Lepus	Leporis	Lep	Hare
Libra	Librae	Lib	Scales
Lupus	Lupi	Lup	Wolf
Lynx	Lyncis	Lyn	Lynx
Lyra	Lyrae	Lyr	Lyre
Mensa	Mensae	Men	Table Mountain
Microscopium	Microscopii	Mic	Microscope
Monoceros	Monocerotis	Mon	Unicorn
Musca	Muscae	Mus	Fly
Norma	Normae	Nor	Set Square
Octans	Octantis	Oct	Octant
Ophiuchus	Ophiuchi	Oph	Serpent Bearer
Orion	Orionis	Ori	Orion
Pavo	Pavonis	Pav	Peacock
Pegasus	Pegasi	Peg	Pegasus
Perseus	Persei	Per	Perseus

Name	Genitive	Abbr.	English name
Phoenix	Phoenicis	Phe	Phoenix
Pictor	Pictoris	Pic	Painter's Easel
Pisces	Piscium	Psc	Fishes
Piscis Austrinus	Piscis Austrini	PsA	Southern Fish
Puppis	Puppis	Pup	Stern
Pyxis	Pyxidis	Pyx	Compass
Reticulum	Reticuli	Ret	Net
Sagitta	Sagittae	Sge	Arrow
Sagittarius	Sagittarii	Sgr	Archer
Scorpius	Scorpii	Sco	Scorpion
Sculptor	Sulptoris	Scu	Sculptor
Scutum	Scuti	Sct	Shield
Serpens	Serpentis	Ser	Serpent
Sextans	Sextantis	Sex	Sextant
Taurus	Tauri	Tau	Bull
Telescopium	Telescopii	Tel	Telescope
Triangulum	Trianguli	Tri	Triangle
Triangulum Australe	Trianguli Australis	TrA	Southern Triangle
Tucana	Tucanae	Tuc	Toucan
Ursa Major	Ursae Majoris	UMa	Great Bear
Ursa Minor	Ursae Minoris	UMi	Lesser Bear
Vela	Velorum	Vel	Sails
Virgo	Virginis	Vir	Virgin
Volans	Volantis	Vol	Flying Fish
Vulpecula	Vulpeculae	Vul	Fox

The Greek Alphabet

α	Alpha	ι	Iota	ρ	Rho
β	Beta	κ	Kappa	σ (ς)	Sigma
γ	Gamma	λ	Lambda	τ	Tau
δ	Delta	μ	Mu	υ	Upsilon
ε	Epsilon	ν	Nu	φ (φ)	Phi
ζ	Zeta	ξ	Xi	χ	Chi
η	Eta	ο	Omicron	ψ	Psi
θ (ϑ)	Theta	π	Pi	ω	Omega

Asterisms

Apart from the constellations (88 of which cover the whole sky), listed on pages 256–258, certain groups of stars, which may form a small part of a larger constellation, are readily recognizable and have been given individual names. These groups are known as *asterisms*, and the most famous (and well-known) is the 'Plough' or 'Big Dipper', the common name for the seven brightest stars in the constellation of Ursa Major, the Great Bear. The names and details of some asterisms mentioned in this book are given in this list.

Some common asterisms

Belt of Orion	δ, ε and ζ Orionis
Big Dipper	α, β, γ, δ, ε, ζ and η Ursae Majoris
Cat's Eyes	λ and υ Scorpii
Circlet	γ, θ, ι, λ and κ Piscium
False Cross	ε and ι Carinae and δ and κ Velorum
Fish Hook	α, β, δ and π Scorpii
Guardians of the Pole	β and γ Ursae Minoris
Head of Cetus	α, γ, ξ², μ and λ Ceti
Head of Draco	β, γ, ξ and ν Draconis
Head of Hydra	δ, ε, ζ, η, ρ and σ Hydrae
Job's Coffin	α, β, γ and δ Delphini
Keystone	ε, ζ, η and π Herculis
Kids	ε, ζ and η Aurigae
Little Dipper	β, γ, η, ζ, ε, δ and α Ursae Minoris
Lozenge	γ, ξ and β Draconis with ι Herculis
Milk Dipper	ζ, τ, σ, φ and λ Sagittarii
Plough	= Big Dipper
Pointers	α and β Ursae Majoris
Pot	= Venus Mirror
Venus' Mirror (or Saucepan)	ι, θ, ζ, ε, δ and η Orionis
Sickle	α, η, γ, ζ, μ and ε Leonis
Southern Pointers	α and β Centauri
Square of Pegasus	α, β and γ Pegasi with α Andromedae
Sword of Orion	θ and ι Orionis
Teapot	γ, ε, δ, λ, φ, σ, τ and ζ Sagittarii
Wain (or Charles' Wain)	= Big Dipper
Water Jar	γ, π, η and ζ Aquarii
Y of Aquarius	= Water Jar

107 Named stars brighter than magnitude 2.75

Name	Con	Mag	Name	Con	Mag
Achernar	α Eri	0.45	**Aspidiske**	ι Car	2.21
Acrab	β Sco	2.56	**Athebyne**	η Dra	2.73
Acrux	α Cru	0.77	**Atria**	α TrA	1.91
Adhara	ε CMa	1.50	**Avior**	ε Car	1.86
Aldebaran	α Tau	0.87	**Bellatrix**	γ Ori	1.64
Alderamin	α Cep	2.45	**Betelgeuse**	α Ori	0.45
Algieba	γ Leo	2.01	**Canopus**	α Car	-0.62
Algol	β Per	2.09	**Capella**	α Aur	0.08
Alhena	γ Gem	1.93	**Caph**	β Cas	2.28
Alioth	ε UMa	1.76	**Castor**	α Gem	1.58
Aljanah	ε Cyg	2.48	**Deneb**	α Cyg	1.25
Alkaid	η UMa	1.85	**Denebola**	β Leo	2.14
Almach	γ And	2.10	**Diphda**	β Cet	2.04
Alnair	α Gru	1.73	**Dschubba**	δ Sco	2.29
Alnilam	ε Ori	1.69	**Dubhe**	α UMa	1.81
Alnitak	ζ Ori	1.74	**Elnath**	β Tau	1.65
Alphard	α Hya	1.99	**Eltanin**	γ Dra	2.24
Alphecca	α CrB	2.22	**Enif**	ε Peg	2.38
Alpheratz	α And	2.07	**Fomalhaut**	α PsA	1.17
Alsephina	δ Vel	1.93	**Gacrux**	γ Cru	1.59
Altair	α Aql	0.76	**Gienah**	γ Crv	2.58
Aludra	η CMa	2.45	**Hadar**	β Cen	0.61
Ankaa	α Phe	2.40	**Hamal**	α Ari	2.01
Antares	α Sco	1.06	**Hassaleh**	ι Aur	2.69
Arcturus	α Boo	-0.05	**Izar**	ε Boo	2.35
Arneb	α Lep	2.58	**Kaus Australis**	ε Sgr	1.79
Ascella	ζ Sgr	2.60	**Kaus Media**	δ Sgr	2.72

Name	Con	Mag	Name	Con	Mag
Kochab	β UMi	2.07	**Procyon**	α CMi	0.40
Kraz	β Crv	2.65	**Rasalhague**	α Oph	2.08
Larawag	ε Sco	2.29	**Regulus**	α Leo	1.36
Lesath	υ Sco	2.70	**Rigel**	β Ori	0.18
Mahasim	ϑ Aur	2.65	**Rigil Kentaurus**	α Cen	-0.29
Markab	α Peg	2.49	**Ruchbah**	δ Cas	2.66
Markeb	κ Vel	2.47	**Sabik**	η Oph	2.43
Menkalinan	β Aur	1.90	**Sadr**	γ Cyg	2.23
Menkar	α Cet	2.54	**Saiph**	κ Ori	2.07
Menkent	ϑ Cen	2.06	**Sargas**	ϑ Sco	1.86
Merak	β UMa	2.34	**Scheat**	β Peg	2.44
Miaplacidus	β Car	1.67	**Schedar**	α Cas	2.24
Mimosa	β Cru	1.25	**Shaula**	λ Sco	1.62
Mintaka	δ Ori	2.25	**Sheratan**	β Ari	2.64
Mirach	β And	2.07	**Sirius**	α CMa	-1.44
Mirfak	α Per	1.79	**Spica**	α Vir	0.98
Mirzam	β CMa	1.98	**Suhail**	λ Vel	2.23
Mizar	ζ UMa	2.23	**Tarazed**	γ Aql	2.72
Muphrid	η Boo	2.68	**Tiaki**	β Gru	2.07
Naos	ζ Pup	2.21	**Unukalhai**	α Ser	2.63
Nunki	σ Sgr	2.05	**Vega**	α Lyr	0.03
Peacock	α Pav	1.94	**Wezen**	δ CMa	1.83
Phact	α Col	2.65	**Yed Prior**	δ Oph	2.73
Phecda	γ UMa	2.41	**Zosma**	δ Leo	2.56
Polaris	α UMi	1.97	**Zubenelgenubi**	α Lib	2.75
Pollux	β Gem	1.16	**Zubeneschamali**	β Lib	2.61
Porrima	γ Vir	2.74			

Further Information

Books

Bone, Neil (1993), *Observer's Handbook: Meteors*, George Philip, London
    & Sky Publishing Corp., Cambridge, Mass.

Cook, J., ed. (1999), *The Hatfield Photographic Lunar Atlas*, Springer-Verlag,
    New York

Dunlop, Storm (2006), *Wild Guide to the Night Sky*, Harper Perennial, New York
    & Smithsonian Press, Washington DC

Dunlop, Storm (2012), *Practical Astronomy*, 2nd edn, Firefly, Buffalo

Dunlop, Storm, Rükl, Antonin & Tirion, Wil (2005), *Collins Atlas of the Night Sky*,
    HarperCollins, London & Smithsonian Press, Washington DC

Grego, Peter (2016), *Moon Observer's Guide*, Firefly, Richmond Hill

Heifetz, Milton D. & Tirion, Wil (2017), *A Walk through the Heavens*, 4th edn,
    Cambridge University Press, Cambridge

Mellinger, Axel & Hoffmann, Susanne (2005), *The New Atlas of the Stars*,
    Firefly, Richmond Hill

O'Meara, Stephen J. (2008), *Observing the Night Sky with Binoculars*,
    Cambridge University Press, Cambridge

Pasachoff, Jay M. (1999), *Peterson Field Guides: Stars and Planets*, 4th edn.,
    Houghton Mifflin, Boston

Ridpath, Ian, ed. (2003), *Oxford Dictionary of Astronomy*, 2nd edn, Oxford
    University Press, Oxford & New York

Ridpath, Ian, ed. (2004), *Norton's Star Atlas*, 20th edn, Pi Press, New York

Ridpath, Ian (2018), *Star Tales*, 2nd edn, Lutterworth Press, Cambridge UK

Ridpath, Ian & Tirion, Wil (2004), *Collins Gem – Stars*, HarperCollins, London

Ridpath, Ian & Tirion, Wil (2017), *Collins Pocket Guide Stars and Planets*, 5th edn,
    HarperCollins, London

Ridpath, Ian & Tirion, Wil (2012), *Monthly Sky Guide*, 10th edn, Dover
    Publications, New York

Rükl, Antonín (1990), *Hamlyn Atlas of the Moon*, Hamlyn, London
    & Astro Media Inc., Milwaukee

Rükl, Antonín (2004), *Atlas of the Moon*, Sky Publishing Corp., Cambridge, Mass.

Scagell, Robin (2014), *Stargazing with a Telescope*, Firefly, Richmond Hill

Scagell, Robin (2015), *Firefly Complete Guide to Stargazing*, Firefly, Richmond Hill

Scagell, Robin & Frydman, David (2014), *Stargazing with Binoculars*, Firefly, Richmond Hill

Sky & Telescope (2017), *Astronomy 2018*, Sky Publishing Corp., Cambridge, Mass.

Stimac, Valerie (2019), *Dark Skies: A Practical Guide to Astrotourism*, Lonely Planet

Tirion, Wil (2011), *Cambridge Star Atlas*, 4th edn, Cambridge University Press, Cambridge

Tirion, Wil & Sinnott, Roger (1999), *Sky Atlas 2000.0*, 2nd edn, Sky Publishing Corp., Cambridge, Mass. & Cambridge University Press, Cambridge

Journals

Astronomy, Astro Media Corp., 21027 Crossroads Circle, P.O. Box 1612, Waukesha, WI 53187-1612.
http://www.astronomy.com

Sky & Telescope, Sky Publishing Corp., Cambridge, MA 02138-1200.
http://www.skyandtelescope.com/

Societies

American Association of Variable Star Observers (AAVSO), 49 Bay State Rd., Cambridge, MA 02138. Although primarily concerned with variable stars, the AAVSO also has a solar section.

American Astronomical Society (AAS), 1667 K Street NW, Suite 800, Washington, DC 20006, New York.
http://aas.org/

American Meteor Society (AMS), Geneseo, New York.
http://www.amsmeteors.org/

Association of Lunar and Planetary Observers (ALPO), ALPO Membership Secretary/Treasurer, P.O. Box 13456, Springfield, IL 62791-3456. An organization concerned with all forms of amateur astronomical observation, not just the Moon and planets, with numerous coordinated observing sections.
http://alpo-astronomy.org/

Astronomical League (AL), 9201 Ward Parkway Suite #100,
Kansas City, MO 64114.
An umbrella organization consisting of over 240 local amateur
astronomical societies across the United States.
https://www.astroleague.org/

British Astronomical Association (BAA), Burlington House, Piccadilly,
London W1J oDU.
The principal British organization (but with a worldwide
membership) for amateur astronomers (with some professional
members), particularly for those interested in carrying out
observational programs.
http://www.britastro.org/

International Meteor Organization (IMO)
An organization coordinating observations of meteors worldwide.
http://www.imo.net/

Royal Astronomical Society of Canada (RASC), 203 – 4920 Dundas Street W.,
Toronto, ON M9A 1B7.
The principal Canadian astronomical organization, with both
professional and amateur members. It has 28 local centres.
http://rasc.ca/

Software

Planetary, Stellar and Lunar Visibility (planetary and eclipse freeware):
Alcyone Software, Germany.
http://www.alcyone.de

Redshift, Redshift-Live.
http://www.redshift-live.com/en/

Starry Night & Starry Night Pro, Sienna Software Inc., Toronto, Canada.
http://www.starrynight.com

Internet sources

There are numerous sites about all aspects of astronomy, and all have
numerous links. Although many amateur sites are excellent, treat any
statements and data with caution. The sites listed below offer accurate
information. Please note that the URLs may change. If so, use a good search
engine, such as Google, to locate the information source.

Information

Astronomical data (inc. eclipses) HM Nautical Almanac Office:
http://astro.ukho.gov.uk

Auroral information Michigan Tech:
http://www.geo.mtu.edu/weather/aurora/

Comets JPL Solar System Dynamics:
http://ssd.jpl.nasa.gov/

Deep-sky objects Saguaro Astronomy Club Database:
http://www.virtualcolony.com/sac/

Eclipses NASA Eclipse Page:
http://eclipse.gsfc.nasa.gov/eclipse.html

Moon (inc. Atlas) Inconstant Moon:
http://www.inconstantmoon.com/

Planets Planetary Fact Sheets:
http://nssdc.gsfc.nasa.gov/planetary/planetfact.html

Satellites (inc. International Space Station)
Heavens Above: http://www.heavens-above.com/
Visual Satellite Observer: http://www.satobs.org/

Star Chart
http://www.skyandtelescope.com/observing/interactive-sky-watching-tools/interactive-sky-chart/

What's Visible
Skyhound: http://www.skyhound.com/sh/skyhound.html
Skyview Cafe: http://www.skyviewcafe.com

Institutes and Organizations

European Space Agency: http://www.esa.int/
International Dark-Sky Association: http://www.darksky.org/
RASC Dark Sky: https://www.rasc.ca/dark-sky-site-designations
Jet Propulsion Laboratory: http://www.jpl.nasa.gov/
Lunar and Planetary Institute: http://www.lpi.usra.edu/
National Aeronautics and Space Administration: http://www.hq.nasa.gov/
Solar Data Analysis Center: http://umbra.gsfc.nasa.gov/
Space Telescope Science Institute: http://www.stsci.edu/

Acknowledgements

16 Wikimedia; 17 Wikimedia; 18 Wikimedia; 19 Wikimedia; 27 NASA/GSFC/SVS; 52 Stephen Pitt; 49 (top) Ian Ridpath; 60 Wikimedia; 61 NASA; 62 NASA; 71 Wikipedia; 72 Natural History Museum, London; 73 National Museum of Natural History, Washington; 95 NASA; 103 AL.com; 110 Peter Komka/EPA-EFE/Shutterstock; 116 Wikipedia; 117 NASA; 125 NASA; 133 Gestrgangleri; 140 Wikipedia; 141 NASA; 149 NASA; 150 NASA; 151 (all) NASA; 166 ESO; 167 ESA; 178 NASA; 179 Dylan O'Donnell; 183 (both) NASA; 197 (both) Wikimedia; 204 NASA; 205 NASA; 206 NASA; 213 (left) NASA; 213 (right) Johan Hagemeyer, Camera Portraits Carme22022ASA; 220 NASA; 227 Wikipedia; 228 (both) NASA.

The authors would also like to thank Barry Hetherington.

Specialist editorial support was provided by Anna Gammon-Ross, Senior Planetarium Astronomer; Affelia Wibisono, Astronomy Education Officer; Luke Hand, Planetarium Astronomer and Julienne Hisole, Assistant Planetarium Astronomer at the Royal Observatory, part of Royal Museums Greenwich.

Index

EXPLORE OUR RANGE OF ASTRONOMY TITLES

Other titles by Storm Dunlop and Wil Tirion

2023 Guide to the Night Sky: Britain and Ireland
978-0-00-839354-0

2023 Guide to the Night Sky: North America
978-0-00-853258-1

2023 Guide to the Night Sky: Southern Hemisphere
978-0-00-853257-4

Latest editions of our bestselling month-by-month guides for exploring the skies. These guides are an easy introduction to astronomy and a useful reference for seasoned stargazers.

Collins Planisphere | 978-0-00-754075-4
Easy-to-use practical tool to help astronomers to identify the constellations and stars every day of the year. For latitude 50°N, suitable for use anywhere in Britain and Ireland, Northern Europe, Canada and Northern USA.

Also available

Astronomy Photographer of the Year: Collection 11
978-0-00-853262-8
Winning and shortlisted images from the 2022 Astronomy Photographer of the Year competition, hosted by the Royal Observatory, Greenwich. The images include aurorae, galaxies, our Moon, our Sun, people and space, planets, comets and asteroids, skyscapes, stars and nebulae.

Stargazing | 978-0-00-819627-1
The prefect manual for beginners to astronomy – introducing the world of telescopes, planets, stars, dark skies and celestial maps.

Moongazing | 978-0-00-830500-0
An in-depth guide for all aspiring astronomers and Moon observers, with detailed Moon maps. Covers the history of lunar exploration and the properties of the Moon, its origin and orbit.

Northern Lights | 978-0-00-846555-1
Discover the incomparable beauty of the Northern Lights with this accessible guide for both aspiring astronomers and seasoned night sky observers alike.

Observing our Solar System | 978-0-00-853261-1
Study the ever-changing face of the Moon, watch the steady march of the planets against the stars, witness the thrill of a meteor shower, or the memory of a once-in-a-generation comet.